Sintering

Monographs in Powder Science and Technology

Series Editor: A. S. Goldberg

Powder Advisory Centre
P.O. Box 78, London NW11 0PG

Roll Pressing—W. Pietsch

Isostatic Pressing—P. Popper

The Production of Metal Powders by Atomization—J. K. Beddow

Sintering—M. B. Waldron and B. L. Daniell

Sintering

M. B. Waldron and B. L. Daniell

University of Surrey, Guildford, Surrey

HEYDEN

London · Philadelphia · Rheine

Heyden & Son Ltd, Spectrum House, Hillview Gardens, London NW4 2JQ.
Heyden & Son Inc, 247 South 41st Street, Philadelphia PA 19104, U.S.A.
Heyden & Son GmbH, Münsterstrasse 22, 4440 Rheine/Westf., Germany.

ISBN 0 85501 178 5

Set by Eta Services (Typesetters) Ltd, Beccles, Suffolk.
Printed in Great Britain by W & J Mackay, Ltd, Chatham.

Contents

Series Foreword vii

Preface ix

Report of Chemical Engineers Task Unit on Sintering x

1 Introduction 1

2 Theory of Sintering 4

3 Theories of Pressure-sintering 13

4 Powder Preparation 16

5 Powder Preparation of Hard Metals 21

6 Process Technology 23

7 Reaction Sintering 27

8 Fibre-Reinforced Materials 33

9 Sintering of Metal and Alloys 44

10 Technology of Pressure-assisted Sintering of Ceramics 50

11 Technology of Pressure-assisted Sintering of Metals 57

12 Equipment (Ceramics) 67

13 Equipment (Metals) 78

References 95

Index 105

Series Foreword

The study and appreciation of particulate materials have advanced markedly in recent years in terms of both interdisciplinary and international activity. Various aspects of powder science and technology have been shown to be of relevance to traditionally separate disciplines, and this field now unites aspects of pharmacy, metallurgy, physical science and many branches of engineering. These extremely valuable areas of advancing technology all too rarely receive the publication they merit due perhaps to paucity of knowledge and the high degree of specialization.

It is the aim of this Series to provide a forum for such work through the medium of concise and timely monographs. We shall publish material collected from a broad spectrum of interest and activity, and promote throughout the interdisciplinary approach and the practical implications. This will, hopefully, catalyse the unification of the overall field by making available up-to-date knowledge in these areas of progress.

London,
April 1976.

A. S. Goldberg

Preface

The membership of the Task Unit set up by the Institution of Chemical Engineers to report on various aspects of Sintering is given on p. x, and their active support throughout the life of the Task Unit and in the preparation of this monograph is sincerely appreciated. The Task Unit commenced its work in 1970 and produced its draft report in 1974. This monograph is an edited version of that report in which account has also been taken of important developments in the last two years.

The editor would like to state that although the individual names are not printed in the text all members of the Task Unit contributed one or more sections of the report. In addition to this the valuable contributions made by Dr D. I. Matkin (Chapter 4), the late Dr L. E. Russell (part of Chapter 6), Professor J. E. Bailey and Dr A. A. Baker (Chapter 8), Mr D. N. Thomas (part of Chapter 9) and Mr R. A. J. Sambell (Chapter 10), are gratefully acknowledged.

This report would never have been completed without the patience and efforts of Mrs J. Rocliffe, Miss M. Etherington and other members of the secretarial staff of the University of Surrey who merit special words of praise.

Guildford
April 1978

M. B. Waldron
B. L. Daniell

Institution of Chemical Engineers Task Unit on Sintering

Chairman

Professor M. B. Waldron, University of Surrey.

Members

Mr J. L. Brackpool, B.S.A. Research Group.

Dr D. W. Budworth, Confederation of British Industry.

Dr B. L. Daniell, University of Surrey.

Mr D. Deverell, AWRE, Aldermaston.

Dr G. M. Fryer, University of Sheffield.

Mr J. D. L. Harrison, U.K.A.E.A., Harwell.

Mr A. J. Hick, Wolfson Heat Treatment Centre, University of Aston in Birmingham.

Mr N. H. Holmes, British Ceramic Research Association.

Mr E. Lardner, Wimet Limited.

Mr J. R. Theobald, Atomic Weapons Research Establishment, Aldermaston.

Mr A. E. White, Morganite Research and Development Limited.

1

Introduction

This monograph attempts to bring within the compass of a modest volume the results of a detailed study undertaken by a Task Unit on Sintering, one of a number set up by the Institution of Chemical Engineers under the general chairmanship of D. Deverell to consider all the aspects of Powder Consolidation. The principal objectives of the Task Unit were to review progress, both of the theory and practice of all aspects of sintering, with the intention of stimulating support for those areas deemed to be the most vital or promising for the further development of powder technology in the United Kingdom. The membership of the Task Unit is given on p. x and their active support throughout the life of the Task Unit and in the preparation of this monograph is sincerely and gratefully acknowledged. The Task Unit commenced its work in 1970 and produced its draft report in 1974. This monograph is an edited version of that report in which account has also been taken of important developments since then.

Sintering is a term used to describe an operation involving a heating process with or without the application of external pressure in which particles are formed into a coherent body. The principal driving force for the reaction is the reduction of surface area. This process can be achieved by solid state reactions or alternatively in the presence of a liquid phase. The three main classes of sintered materials are: sintered metals, sintered ceramic bodies and sintered ore bodies. In most cases, after sintering, the product has some form of geometrical shape (bushes, gearwheels, filters, etc.) and in the case of metals considerable economy is achieved by the avoidance of machining operations and by lower processing temperatures compared with melting and casting although offset by higher materials costs. The technique also permits special features to be incorporated in the products such as controlled porosity as in metal filters, or the distribution of a second constituent as in cermets.

The simplest case of sintering is that of a pure metal which is normally effected at temperatures of about 0·5 of the melting point.

In the early stages of sintering the most significant effect is a change of pore shape, without necessarily much change in pore volume, from an irregular form which is characteristic of the as-pressed condition towards a spherical form which is characteristic of the sintered condition. The driving force for this is clearly a reduction of surface energy.

In this stage of sintering increases in strength can be obtained by the use of activated sintering which normally involves some addition, either to the atmosphere or to the compact used to form an unstable volatile metal compact, which continuously forms and is dissociated, thereby forming a clean, very active surface or a transient liquid phase. It has been suggested that the normal oxide films which are found on all metal powders fill the former role.

To obtain a 100% density a technique which has been used for many years is that of liquid phase sintering, typically of the tungsten carbide–cobalt mixtures, in which the sintering operation is carried out above the pseudo-binary, tungsten carbide–cobalt eutectic. In this case effectively theoretical densities can be obtained in extraordinarily short times. Liquid phase sintering is also important in the firing of refractory materials, but in this case the liquid is formed not only from deliberate additions but also from 'impurities' which exist in the bulk raw material.

Liquid phase sintering is advantageous in a sense, but quite clearly it means that sintering is carried out at a temperature well below the melting point of the more refractory phase and usually limits the temperature of use. In metals this is normally of no great importance, but in refractory materials and to obtain a high density in these materials this technique is replaced by a hot pressing.

The monograph therefore includes reference to hot pressing or, as it is sometimes called, pressure-assisted sintering, including the technique of Isostatic Pressing which has recently become of commercial importance in the manufacture of large, pore-free parts in cemented carbides and other higher-temperature materials.

In general, sintering theories are still of only qualitative value to the industry, and often only confirm trends that are already known empirically. The main difficulty in making qualitative predictions is that metal powders do not resemble the model systems for which the theories have been derived.

A slightly different method is reaction sintering. There are three general cases. In the simplest case it can be applied to the technique in which a powder which decomposes on heating to form the final compound is substituted for the normal powder in the conventional sintering operation. In the other cases two or more components of the required ceramic compound react together during the sintering operation. In the first of these cases the two components in powder form are mechanically mixed, shaped and reaction-sintered together. In the second case where one of the compounds is gaseous at room temperature (oxides or nitrides) or has a low melting point relative to the other constituent, e.g. phosphides, sulphides, silicides or aluminides, the more refractory component in powder form is shaped and reacted with the other constituent into a gaseous or liquid form.

Two cases of agglomeration are different from most others. Iron ore, which is sometimes specially ground, at other times is the residue of other crushing processes or alternatively is the product of mineral dressing operations, is agglomerated by being combusted on a strand with coke additions made to it. This material, referred to as sinter, is produced in the form of irregular lumps which can be handled and charged into a blast furnace. Another special case is the formation of carbon refractories. In many carbon products the ground coke, for example, is bonded with tar and heated in a reducing atmosphere to a temperature of approximately 1000°C. This will coke the tar and at the end provide a carbon body of high strength and volume stability at temperatures well in excess of those at which the operation was carried out.

The success of sintering is frequently judged, in the case of metals, by mechanical properties, the reduction of area, the ultimate tensile strength and the density; or in the case of ceramics, by the porosity, cold crushing strength and permanent changes on re-heating and similar parameters, described in BSS 1902 (1967), all of which give a guide to the degree of success of the sintering operation. In the case of refractories and ceramics any permanent changes on re-heating should be minimal. In both metals and ceramics, sintering is carried out on a body which has first been shaped to form a green compact, and it is axiomatic that a successfully sintered body cannot be produced unless the preparation stage has been properly carried out. Recent improvements in technology have resulted from greater attention being paid to the preparation and characterization of powders because of their influence on sintering. Milling treatments, which determine the particle shape, size and size distribution and also condition the particle surfaces and induce strain energy, will be very significant.

This monograph will therefore include sections on sintering theory, powder preparation, sintering technology, special techniques and reference to the furnaces in which sintering is carried out. Pressing and similar operations are referred to elsewhere.[1–5]

2

Theory of Sintering

INTRODUCTION

The term sintering is used in several different contexts with meanings which differ somewhat from each other. The common feature is that all processes to which the name is applied are such that diminution of the surface area of an assembly of particles occurs in the process, usually under the influence of heat, and the diminution of the surface energy which accompanies the loss of area is the driving force for the process. Some strengthening of the aggregate usually accompanies the loss of surface area, and is almost always a desired feature of the process: a degree of densification is also usual although expansion can occur. In some contexts, such as the sintering of many ceramic bodies, the shrinkage which results from the densification is accepted as an inevitable concomitant of this convenient method of preparing solid articles, but in the powder metallurgy field the main aim is to produce precisely dimensioned parts of considerable complexity by an economic method, and efforts are made to minimize the dimensional changes whilst still obtaining strengthening of the article during sintering. In other applications, such as the manufacture of, for example, filters, catalyst supports, and feed materials for chemical processing, the aim is to produce a highly porous but coherent finished product from the initial powder, and densification is avoided or prevented as far as possible.

GEOMETRY OF SINTERING

Since the diminution of the surface energy of the assembly of original particles is the dominating feature in sintering, the geometrical arrangement of this surface largely determines what can happen during sintering although not whether or how fast it will happen. On the whole, most attention has been focused on the kinetics and mechanisms of sintering and rather less on the geometry of the process. The geometry of powder compacts is necessarily rather complex, and the subject is not well understood in detail, particularly for the early stages of sintering.

In general a multiphase body consists of a series of cells or grains which meet, in stable configurations, three to an edge and four to a point. These three-grain edges and four-grain corners, as they are known, are the fundamental units of structure, and their shapes are determined by the equilibrium between the various surface energies involved.

When there are two phases present (of which one may be porosity) the surface energies involved are those of the interfaces between one phase and the other and between two grains of one phase. For the case of most interest where only one of the phases is capable of having a grain boundary the second phase must meet such a boundary at a fixed *dihedral angle* ϕ and it is easily seen (Fig. 1) that

$$\cos\frac{\phi}{2} = \frac{\gamma_{AA}}{2\gamma_{AB}} \tag{1}$$

A plot of ϕ against γ_{AA}/γ_{AB} is shown as Fig. 2, from which it can be seen that the dihedral angle is very sensitive to the ratio γ_{AA}/γ_{AB} when the ratio is near 2. Values in this region are often found for compositions of interest to ceramicists. Grain boundary energies are usually somewhat less than half the surface energy of the material in which they are situated, so that the dihedral angle of a pore tends to be in the 150–180° range, and the equilibrium geometry of a solid plus pores may thus differ considerably from that of a solid plus liquid.

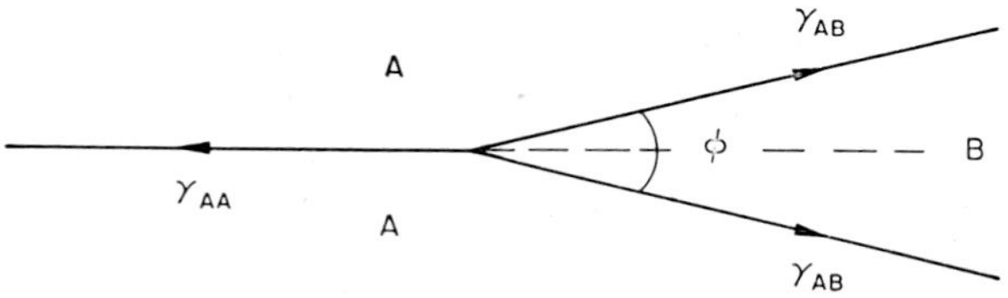

Fig. 1. Schematic diagram of a two-phase system.

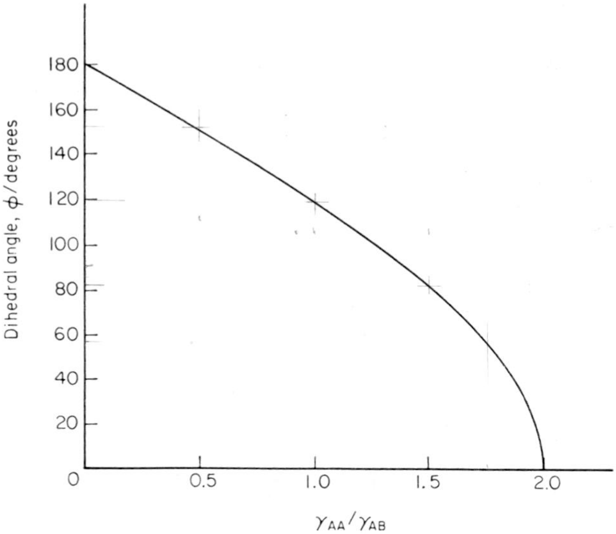

Fig. 2. The variation of dihedral angle with the ratio of surface energies.

When a second phase, either porosity or a liquid, is present, it is energetically favourable for it to occupy, in order of preference, four-grain corners, three-grain edges, (two-grain) faces, and lastly sites in the interior of grains, because the total amount of interface, and hence of interfacial energy, is thereby reduced. The lower the dihedral angle, the more energetically favourable it is for the second phase to remain at a boundary.

It can be shown by geometrical arguments that at dihedral angles of 60° or less, penetration of the second phase occurs along the three-grain edges of the first phase, producing a structure consisting of two interlinked continuous phases, but that penetration between grain faces only occurs when the dihedral angle falls to zero.

For a full account of the above topics, reference should be made to Smith.[6]

THE STAGES OF SINTERING

For a collection of particles of a single phase the dihedral angle will usually be high and therefore points of contact between particles will grow into necks if some mechanism is available to allow the equilibrium geometry to be reached. This stage of neck growth is known as the first stage of sintering. After a time, and at a point where the porosity is about 15%, the grain-boundary energy begins to be a significant contributor to the total energy of the system and the grain boundaries begin to rearrange themselves to minimize their total area. The geometry in this second stage of sintering thus becomes that of an assembly of polyhedral grains with pores along the three-grain edges, and the tendency to minimize the grain-boundary area results in grain growth. The arrangement of the grain boundaries is very similar to that of the soap films in a froth, and it can be shown that the average cell or grain possesses 13·4 faces, 22·8 vertices, and 34·2 edges.[6,7] The pores continue to shrink as sintering proceeds and in the third stage of sintering they become unstable as approximate cylinders along the three-grain edges and pinch off to become isolated pores at four-grain corners.[7,8] This stage commences at about 5% total porosity and may continue until all porosity is eliminated. Typical curves of open and closed porosity are shown in Fig. 3.

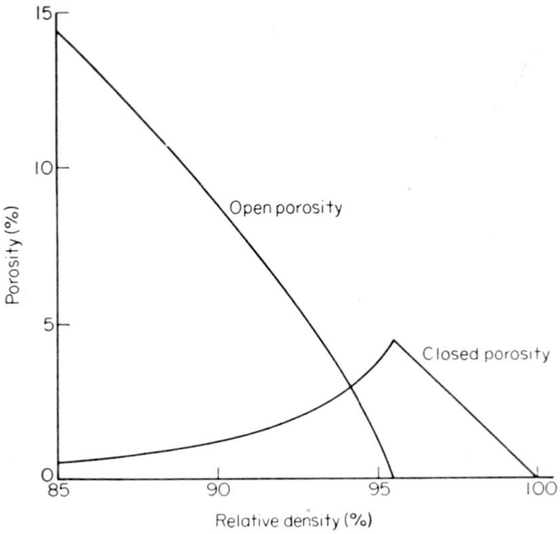

Fig. 3. The change of open and closed porosity with relative density.

A new phenomenon may occur during the third stage of sintering. If gas is trapped in the closed pores its pressure will rise as the pore radius r decreases until it may prevent further shrinkage when the equilibrium condition $p = 2\gamma/r$ is reached, γ being the surface energy. It is, however, energetically favourable from the point of view of energy stored in the compressed gas, and neutral from the point of view of surface energy, for gas to transfer from small pores at high pressure to large pores at low pressure. If such transfer is possible, the result is an increase in the total volume of the piece which is described as *bloating*. A theory of the kinetics of this process yields the relation:[9]

$$r_t^2 - r_0^2 = At \tag{2}$$

where r is the pore radius, t is time, and A is a constant which depends linearly on the solubility of the gas and on its diffusion coefficient.

For a system in which liquid is present, the dihedral angle may be low, so that when the liquid forms on initial heating it penetrates between the solid particles to a considerable extent. Spherical pores form in the liquid and the reduction of the surface energy of these pores by shrinkage draws the particles together. Relative movement of the particles is facilitated by the liquid and so they can pack quite efficiently, and a solid skeleton of tightly packed particles is quickly formed. The tendency for pores to amalgamate is still present and this process may be easier than in the absence of the liquid.

GRAIN GROWTH

The dihedral angle determines the overall distribution of phases but, as has been mentioned above, the total energy of the system can be decreased by increasing the scale of the texture, and thus reducing the amount of interfacial area in unit volume. The resulting grain growth is an essential feature of the sintering process and one which complicates its understanding and control. Grain growth in an aggregate takes place with some grains growing and some shrinking, as the total number of grains decreases.

Grain growth can take place readily in a single-phase material because it only involves small local movements of atoms, but the presence of small amounts of second phase may slow the process considerably because its constraint to grain boundaries of the first phase means that it has to be moved *en masse* for grain growth to occur. Zener[10] has shown that strongly curved grain boundaries will be able to pull free of inclusions but that as the grains grow and their boundary curvature decreases, they will become pinned when:

$$R = \frac{4r}{3f} \tag{3}$$

where R is the radius of curvature of a boundary, r is the radius of the inclusions, and f is their volume fraction.

It is often assumed that $R \sim D$, the grain size, but observation suggests[11] that a more realistic value is $R = 10D$.

Grain growth in the pure single-phase state is often assumed to follow the law:

$$D_t^2 - D_0^2 = Bt \tag{4}$$

but observations tend to be more consistent with the relationship:

$$D_t^3 - D_0^3 = Ct \tag{5}$$

which has been explained as due to the drag imposed by pores which act as slowly mobile second-phase particles.[12,13]

It is possible for pore growth to occur by this mechanism without gas being transferred through the solid by the coalescence of pores at the corners of a small grain which disappears during grain growth.

If a large amount of second phase is present, then grain growth in the first phase always appears to follow the cube law given above, and this law has good theoretical backing in terms of solution and reprecipitation under the influence of size effects on solubility.[14–16]

MECHANISMS OF MATTER TRANSPORT

The adjustments needed to minimize the surface energy in a body undergoing sintering necessarily involve the movement of matter. In the case of a single solid phase, matter can be moved under the influence of surface energy from the neighbourhood of convex surfaces to that of concave surfaces by several mechanisms. These are:

(i) evaporation and subsequent recondensation, under the influence of surface curvature on vapour pressure;
(ii) diffusion over the surface, atom by atom;
(iii) plastic flow by dislocation movement in a crystalline material;
(iv) viscous flow in an amorphous material, particularly in a liquid phase;
(v) bulk diffusion through the solid, down the vacancy gradient produced by the pressure gradient resulting from the surface curvature variations.

Of these processes, the first two are not capable of moving the centres of particles closer together, and hence of causing densification, although they can cause neck growth and hence strengthening of the initial assembly or compact of particles. The other processes can cause both neck growth and shrinkage.

Studies of the mechanisms of sintering have been made by using simple systems which can be treated theoretically and also observed in practice. A favourite system is that of a sphere of radius r on a plate, for which it has been shown[17,18] that the neck radius x should obey the law:

$$(x/r)^n = kt/r^m \tag{6}$$

where t is time and k is constant. n and m are constants which vary with the mechanism of matter transport as shown in Table 1.

Studies of neck growth, necessarily made on spheres which are rather larger than powder particles in normal use, have shown that glass spheres sinter by viscous flow, but that crystalline metals and ceramics usually sinter by bulk diffusion.[19] There is evidence that flow mechanisms contribute to the early stages of neck formation.[20]

The discrimination between mechanisms is, however, not very clear and misleading inferences about the behaviour of compacts have sometimes been drawn from studies of model systems.

TABLE 1.

Mechanism	*n*	*m*
Viscous or plastic flow	2	1
Evaporation/condensation	3	2
Bulk diffusion	5	3
Surface diffusion	7	4

Neck growth studies are obviously relevant to the first stage of sintering, and theories concerned with the second and third stages are somewhat more difficult, particularly because of the difficulties introduced by grain growth. A most important experimental observation[21] which has been incorporated into the theories is that pores shrink only when they are situated on grain boundaries and that if they become isolated in the centre of a grain they cease to shrink. The usual interpretation of this observation is that grain boundaries act as a sink for the vacancies of which the pores are composed, and that bulk diffusion is the mechanism of matter transport. Theories of the kinetics of densification have been worked out for the second and third stages of sintering on this basis,[22] and agree reasonably well with experiment, but there are some difficulties in achieving complete numerical agreement.

The equations derived are of the form:

$$\frac{dP}{dt} = -C\frac{D\gamma v}{l^3 kT} \tag{7}$$

where P is porosity, t is time, D is a diffusion coefficient, v the volume of a vacancy, γ the surface energy, l is a measure of grain size, and C is a constant. The difficulty has been in relating the values of the constants to the geometry of a real compact. This theory predicts that sintering will continue until all porosity is removed in a finite time, and experiment has shown that this state can be achieved if certain precautions are taken.

A rather more sophisticated model which allows for surface, grain boundary and bulk diffusion has recently been put forward[23] for the second stage of sintering. By suitable computer programs, the contribution of the separate mechanisms, even when operating simultaneously, can be evaluated, for well defined neck geometries only.

An early theory of third-stage sintering was put forward by Mackenzie and Shuttleworth,[24] who considered a homogeneous body containing spherical pores densifying by plastic or viscous flow under the influence of the surface energy of the pores. For the viscous case, they derived the equation:

$$\frac{d\rho}{dt} = C\frac{\gamma n^{1/3}}{\eta}(1-\rho)^{2/3}\rho^{1/3} \tag{8}$$

where ρ is the relative density, γ is the surface energy, η is the viscosity, and n the number of pores in unit volume. The constant C is $3/2(4\pi/3)^{1/3}$. More complex equations were derived for plastic flow of the Bingham type in which the material exhibits a yield point. A significant feature of this theory is that it predicts that sintering in a material which exhibits a yield point will terminate at a stage where porosity remains.

Recently Ashby[25] has developed the concept of sintering diagrams. For specific single-component systems of known geometry they show the dominant mechanism of sintering and the net rate of neck growth or densification for a given temperature and neck size. The diagrams are constructed using existing equations from the literature and therefore retain their inherent limitations. However, the diagrams indicate the dominant mechanisms in a given situation and enable conclusions to be drawn as to the effect of experimental or production variables.

When a liquid phase is present, different considerations apply. In particular, the effect of curvature, and hence of particle size, on the solubility of the solid in the liquid becomes important in a way which is formally equivalent to the dependence of vapour pressure on curvature. This provides a mechanism for the initial rounding-off of particles and leads to the achievement of good packing, and subsequently provides a mechanism for grain growth in two-phase systems. Early attempts to provide theories for liquid phase sintering (known as 'vitrification' in the ceramic industry) relied on the assumptions of complete wetting of the solid by the liquid, i.e. zero dihedral angle, and some solubility of the solid in the liquid. Such was the 'heavy alloy theory' of Cannon and Lenel[26] and also Kingery's[27] approach. These theories allow for an initial melting stage and rearrangement of particles, followed by a solution and reprecipitation stage, with a final stage in which a rigid skeleton gradually densifies by solid state processes, although this should only occur if the dihedral angle is finite.

The most commercially important 'hard metal' system, that based on tungsten carbide bonded with cobalt, is an almost perfect example of a zero dihedral angle liquid–solid system exhibiting appreciable solubility, once the eutectic temperature of about 1285°C is exceeded. The liquid penetrates very rapidly down grain boundaries in the solid and complete densification occurs in a few seconds as the pores in the liquid phase are eliminated.

Recent studies of ceramic systems with finite dihedral angles, which appear to be more generally representative conditions, have shown that the angle exercises considerable influence on the densification and grain growth in the system[28,29] but no comprehensive theory has been published. An analysis of the capillary forces affecting the liquid-phase sintering of jagged particles has, however, appeared.[30] Reviews of present theories of liquid-phase sintering are available.[31,32]

THE RELATION OF THEORY TO EXPERIMENT

The relationship between theory and experiment in sintering is qualitatively good but in some cases quantitatively bad, largely because the geometry of real compacts and particularly of real powder particles is rather more complex than can be conveniently handled mathematically.

In general terms, theoretical considerations show that sintering can be promoted by introducing a liquid phase or by reducing the particle size and hence increasing the driving force for sintering. Both these expedients are well known and much employed. Since most of the matter transport must take place near the surfaces of particles, the incorporation of carrier phases on the surfaces, or specially mobile surfaces, is also practised. If, on the other hand, shrinkage must be minimized and porosity retained, theory shows that large particles will be helpful, as will the promotion of evaporation/condensation and surface diffusion. A more sophisticated way of achieving the same effect is to pin the grain boundaries by means of small amounts of second phase, thus largely immobilizing the microstructure. At least one application of this idea has been patented.[33] In the case of metal components in which high density and strength are required but shrinkage must be small, the usual technique is to form the initial powder into a high density compact by pressing at a high pressure, and thus limiting the scope for further densification. This process is not possible for most ceramics, as they will not deform sufficiently to pack to much better than about 60% relative density, whereas 85% density can be taken as a representative figure for metals.

A good illustration of the use of theory to guide experiment is the development of pore-free sintered materials.[34–36] It is normally found that when only a little porosity (1–2%) is left, grain boundaries begin to pull away from pores because the total energy of the local system is lowered by the straightening of a curved boundary more than it is increased by the separation of the pore and the boundary. The isolated pore then ceases to shrink. By the use of grain growth control agents, this behaviour can be suppressed, resulting in sintering proceeding until all porosity is removed, provided that the sintering is performed in an atmosphere of a gas which does not become trapped in the pores and prevent their shrinkage in the manner described above.

The main discrepancies between theory and experiment arise because real powders depart very markedly from the ideal spherical form usually assumed. This is well illustrated by their packing properties. A random packing of uniform spheres packs to about 60% density, whereas real powders may pack to less than half this value unless pressed at very high pressure. This behaviour arises from a number of causes, but principally because of aggregation of the particles into clusters which then pack together loosely. The resultant is a double porosity, one part inside the clusters and one part between them. This structure can be such that some particles make insufficient contact with others for necks to form, so that sintering does not take place in some regions. A considerable amount of work has been done on the preparation of powders for sintering, and much more remains to be done, for it is still found that the variation of behaviour between powders of the same material is very great, and this aspect of sintering is of greater importance to the manufacturer of sintered articles than the exact kinetics or mechanism of the sintering process.

FUTURE WORK

Attempts to elaborate theories of sintering on the lines pioneered by Kuczynski[17] and Coble and Burke[19] do not appear to have added much to our knowledge of the sintering process, and the emphasis in these theories on the determination of the mechanism by which sintering occurs has led to little work of practical value. The emphasis in sintering theory is therefore changing, as a demand for useful information begins to make itself felt.

Areas in which progress appears to be possible are the geometry of less-idealized systems, in particular in the treatment of grain growth and of interactions between pores and grain boundaries; and the theory of sintering of two-phase systems, in which the geometrical concept of dihedral angle provides a link between observable phenomena and the balance of physical forces which give rise to them.

In many cases, the features of greatest significance in a practical context, such as impurities and atmosphere, are specific to the particular system being studied, and the prospects for any general theoretical development do not appear to be very good.[37–39]

3

Theories of Pressure-sintering

When a powder compact is sintered under the application of an external pressure the initial stage of compaction, probably up to a relative density of about 0·85, is a complex process of particle packing, sliding, fragmentation and deformation, and is not likely to yield to theoretical analysis. The subsequent intermediate (connected porosity) and final (closed porosity) stages both involve a solid matrix with a definite pore system, and should be amenable to theoretical treatment, although only limited success has so far been achieved. A successful pressure-sintering theory would be valuable because (a) it would enable experimental data for a given material to be extrapolated to predict results under changed conditions, and (b) it would enable viscosity or diffusion data to be calculated, and so afford a means of assessing the results of changing the composition of the material being studied. In the present survey attention will be concentrated on theories which lead to practical densification equations which can be used by workers seeking a mathematical fit to experimental data.

No real contribution to the theory of pressure-sintering was made until after the publication of the Mackenzie–Shuttleworth theory of (pressure-less) sintering.[24] The model used in this theory comprised a homogeneous body, deformable by viscous or Bingham flow, containing spherical pores. The material around each pore was assumed to flow radially towards it (under the influence of the surface energy of the pore), the work done at the pore surface being equated to that expended in deforming the surrounding material. As was pointed out by Murray *et al.*,[40] the theory could be extended to cover the application of an externally applied pressure. This resulted in the first published theory of pressure-sintering, represented by the equation:

$$\frac{\mathrm{d}\rho}{\mathrm{d}t}\left(=-\frac{\mathrm{d}P}{\mathrm{d}t}\right)=\frac{3}{4}\frac{P}{\eta_\infty}\left[\sigma-\sqrt{2}\,\tau\ln(l/P)\right] \tag{9}$$

where ρ is the relative density, P the fractional porosity ($P = 1-\rho$), σ the applied pressure, τ the critical shear stress of the material, and η_∞ is the viscosity at infinite strain rate. For a viscous material ($\tau = 0$) the equation becomes

$$\frac{\mathrm{d}\rho}{\mathrm{d}t}=\frac{3}{4}\frac{P}{\eta}\sigma \tag{10}$$

It is noteworthy that these equations do not depend on the pore separation. In Eqns (9) and (10) it is assumed that the applied pressure is great compared with the surface-energy forces at the pores: if this is not so the pore force $2\gamma/r$, where γ is the surface energy and r the pore radius, should be added to σ in the equations. When integrated Eqn (10) becomes

$$\ln P = -\frac{3}{4}\frac{\sigma}{\eta}t + \text{constant} \tag{11}$$

so that a plot of ln P versus t should yield a straight line.

Equation (10) was convincingly verified for silica by Vasilos.[41] It was also used by Mangson *et al.*[42] for analysing results using alumina, and by Jaeger and Egerton[43] for results with potassium–sodium niobates. These latter workers, however, did not directly verify Eqn (11).

It was later found that Eqns (9) and (10) did not describe data obtained with beryllia (McClelland[44]) or with alumina or magnesia (Vasilos and Spriggs[45]), which is not surprising as it is unlikely that these materials will deform as viscous or Bingham solids. McClelland obtained a better fit to his data by multiplying the applied pressure by a factor $(1-P^{2/3})^{-1}$, but by doing so departed from the concepts of the original theory. Koval'chenko and Samsonov[46] derived an alternative equation, based once again on pores in an incompressible viscous medium, and produced an equation not unlike Eqn (10):

$$\frac{d\rho}{dt} = \frac{\sigma P}{4\eta}\frac{(3-P)}{(1-2P)} \tag{12}$$

At this stage some confusion appears to have arisen. Pressure-sintering finds its greatest use with ceramic materials, and unfortunately most ceramics do not deform as viscous or Bingham solids. In searching for a means of describing the pressure-sintering behaviour of materials which deform by vacancy diffusion, various workers were influenced by a paper *Diffusional Viscosity of a Polycrystalline Solid* by Herring.[47] In this Herring derived an equation for the strain rate of a polycrystal subjected to a uniaxial stress σ:

$$\dot{\varepsilon} = \frac{KD\Omega\sigma}{l^2kT} \tag{13}$$

where $\dot{\varepsilon}$ is the strain rate, K a numerical constant, D the vacancy diffusion coefficient, Ω the vacancy volume, l the grain size, k Boltzmann's constant and T the absolute temperature. Unfortunately, he also gave the 'effective viscosity' of the polycrystal

$$\eta = l^2kT/KD\Omega \tag{14}$$

which was really no more than the viscosity of a hypothetical viscous material which would deform at the same rate. This led some workers to try to modify the viscous-flow pressure-sintering theories by introducing the Herring 'viscosity' relation. Thus the final stage of the derivation of the Koval'chenko and Samsonov model involved the substitution of the Herring 'viscosity' for η, and a correction for grain growth. Scholz and Lersmacher[48] pointed out that the resulting equation could be cast into the form

$$P = P_0(l+bt)^{-n} \tag{15}$$

where b is a constant arising from the grain-growth correction and n depends on various properties of the material. $P = P_0$ when $t = 0$.

The Herring creep model was also used to try to interpret pressure-sintering results for alumina and magnesia by Vasilos and Spriggs[45] and for alumina by Rossi and Fulrath,[49] the latter authors attempting to show that the form of Eqn (10) was equally applicable to either viscous-flow or diffusion-controlled pressure-sintering.

Scholz[50] found that Eqn (15), regarded as an empirical equation, was useful in interpreting data obtained when hot-pressing metallic carbides. In using the equation, b is determined from a plot of $\ln P$ versus $\ln t$: when the curve at high values of t is extrapolated as a straight line back to $\ln P_0$, the resulting value of time t_s enables b to be found from $bt_s = 1$. $\ln P$ may then be plotted against $\ln(1+bt)$ to give a straight line. It is worth noting that Eqn (15) may be differentiated to give

$$\frac{dP}{dt} = -nbP_0{}^{-1/n}\rho^{n+1/n}$$

so that it can represent any theory where the densification rate is proportional to the porosity raised to a power greater than unity.

Clearly, any truly successful theory of pressure-sintering by diffusion must be derived from a model which uses the concepts of deformation by diffusion as its only basis. If a porous polycrystal containing thermodynamic vacancies is subjected to an external pressure the normal stress at the grain boundaries will cause the vacancy concentration to be depressed in their vicinity.[51] There will be no normal stress at the pore surfaces, other than a negative stress caused by their surface energy which will tend to increase the vacancy concentration in this position. As a result of these effects a vacancy concentration gradient will exist between pores and grain boundaries, and vacancies will flow from the former to the latter. The grain boundaries will collapse and the pores will shrink.

Using similar concepts to these, Fryer[52,53] derived an equation for final-stage (closed pore) sintering:

$$\frac{d}{dt}\left(\frac{P}{\rho}\right) = -z\frac{\sigma D\Omega}{l^2kT}\left(\frac{P}{\rho}\right)^{5/3} \tag{16}$$

where z is a constant. When integrated this becomes:

$$\left(\frac{\rho}{P}\right)^{2/3} = \frac{2}{3}\left(\frac{z}{l^2}\right)\frac{D\Omega\sigma t}{kT} + \text{constant}$$

so that a plot of $(\rho/P)^{2/3}$ versus t should give a straight line. The equation was found to work very well with data for the pressure-sintering of alumina, at least between the relative densities 0·87 and 0·96. This is really the intermediate range of densities when porosity is continuous, and not the final stage for which the equation was derived. However, there were various errors in the theoretical derivation, and Eqn (16) should be regarded as semi-empirical. It is most unlikely that alumina can deform by other than a diffusion mechanism at the pressures and temperatures used, so that Eqn (16) is probably the most useful so far available for describing diffusion-controlled pressure-sintering.

4

Powder Preparation

INTRODUCTION

It is well established that the characteristics of starting powders strongly influence all stages of ceramic fabrication,[54] and hence influence the microstructure and thus the physical properties of the final ceramic body.[55] Powder characteristics are determined by the initial powder preparation and by the subsequent powder treatment.

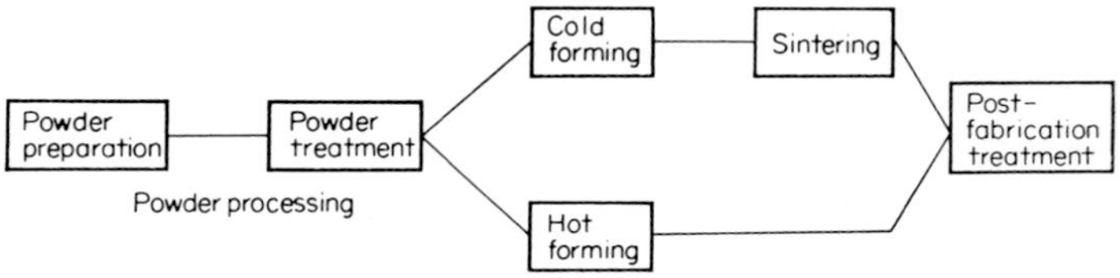

Fig. 4. Conventional fabrication routes for ceramics.

The total fabrication process can be represented diagrammatically as in Fig. 4; it can be seen that powder processing is an important stage of any fabrication route, and yet since it contains the greatest number of variables it is the most difficult to characterize and control. The characteristics of a powder, at any stage of processing, can be described by a large number of parameters, which have been reclassified by Stover[56] and are summarized in Table 2. It is important to emphasize that at most stages of processing, a powder will consist of individual crystallites that are bonded together to

TABLE 2. Important powder characteristics

Crystallite	Crystallographic form, Stoichiometry Size, Size distribution Shape Impurity content and distribution (including adsorbed gases on surface) Activity, Reactivity of surface energy Internal energy
Particle Agglomerate Aggregate Granule	Size, Size distribution Shape State of agglomeration Pore structure Impurity content and distribution Powder 'stickiness' or 'glidability'

form a particle which may be described as an aggregate, an agglomerate, or a granule. Hence the characteristics of both crystallite and particle are important. The major aim of powder processing is to obtain a reproducible powder with a set of known and required characteristics, in order to predict the behaviour of the material during the subsequent fabrication stages.

POWDER PREPARATION

Ceramic powders can be prepared by comminution of bulk material, by chemical preparation or in a few specialized cases directly by vapour phase reactions.

Comminution (Crushing and Coarse Grinding)

Raw materials for the more traditional ceramics receive minimal processing after mining. The naturally occurring minerals can be reduced to coarse powder by a series of different comminution processes including gyratory crushing, jaw crushing, roll crushing and hammer milling. The coarse powders may be subsequently processed by mineral dressing techniques.

Tonnage quantities of Al_2O_3, SiC, and BC_4 are produced by fusion techniques and the resultant fused lumps are broken down to coarse powder by various comminution processes. Williams[54] has discussed the main features of powders prepared by comminution, and emphasizes that the main limitations are control of particle shape and size, and control of physical and chemical properties of powder surfaces.

Chemical Preparation

The most common chemical preparation route for ceramic oxide powders is the thermal decomposition of metal salts or hydroxides. Several workers have studied the characteristics of single oxide powders produced by calcination (the reader is referred to reviews on Al_2O_3,[57] Al_2O_3 and MgO,[58] and BeO[59]) and have shown that the properties of a sintered body can be strongly influenced by the crystallite size, aggregate size and morphology and inter-aggregate porosity of the powder from which it is prepared. Methods of chemical precipitation and controlled evaporation of solutions have been developed to yield the powders or their precursors in the optimum physical forms.[60–64]

Ceramic powders of mixed oxides, such as ferrites, can be prepared by solid state reaction between constituents that have been co-precipitated or physically mixed. Non-oxide powders can also be prepared by reaction between constituents. For example, metal borides can be prepared by a wide variety of solid state reactions,[60] while silicon nitride powder can be prepared by reaction between silicon powder and nitrogen gas. It is difficult to control the aggregate characteristics of both single and mixed oxide powders prepared by these techniques; indeed their complex nature makes characterization difficult. Such powders frequently require treatment prior to the forming and sintering stages of the fabrication route.

POWDER TREATMENT

Most ceramic powders require some further treatment after preparation prior to the forming and sintering stages of fabrication. The powder treatment may be designed primarily to influence the flow properties or packing properties of the powder, or designed primarily to enhance the densification during sintering.

Comminution (Fine-grinding)

This may be considered as having the following stages:

(a) breaking down of aggregates to single crystallites;
(b) fracture of individual crystallites and production of defects; and changes within the crystallites by longer milling;
(c) re-agglomeration at long milling times.

The above stages are clearly relevant to the treatment of a chemically prepared powder containing aggregates, although any fine grinding of crushed raw materials will be an extension of the early comminution stages. The strength of the as-prepared aggregates of chemically prepared powders is very important during stage (a), and has been studied for a number of Al_2O_3 powders by Nietz *et al.*[65] The sequence of stages of ball-milling of powders has been studied by Harwood *et al.*[66] for Fe_2O_3, by Denton *et al.*[67] for BeO, by Gitzen[57] for Al_2O_3. Denton *et al.*[67] observed during stage (b) that there was not only a gradual reduction in average crystallite size, but also a reduction in the range of crystallite size distribution of the BeO powder. The all-important question of the ultimate fineness that can be achieved by comminution has been discussed by Meloy.[68] Wet-grinding has been the accepted technique until quite recently; however dry-grinding averts the use of a de-watering stage, and Hart and Hudson reported that organic grinding aids allowed efficient dry-grinding of Al_2O_3 without packing of the powder within the ball-mill.[69]

During stage (b) it is possible that in addition to a reduction in crystallite size, plastic deformation or even phase changes may occur within individual crystallites. The origin, nature and effects of crystallite strain has been reviewed by Hirschhorn.[70] More specific studies of X-ray line broadening have been made by Chiu and Day[71] on a number of commercial powders (including dry-ground commercial Al_2O_3 powder), and by Lewis and Lindley[72] and Wheeler and Lewis[73] and on Al_2O_3 powders after milling. Chiu and Day, Lewis and Lindley and Lewis and Wheeler observed that the strain increased with milling time to a saturation value of around 10^{-3}, and could be annealed out after a few hours at 1200°C. It is therefore uncertain whether this strain will contribute to the sintering activity of milled powders. Phase transformations have been observed in calcite and lead oxide after extensive milling.[74]

A final comment should be made about contamination of the powder by pick-up from the milling media. If steel milling media are used then iron and soluble impurities can be partially removed by acid leaching. However the increasing demand for high purity oxides, particularly Al_2O_3, has led to a plea for 99% alumina mill-liners and grinding media.[75]

Blending

A wide range of different additives requires to be made to a ceramic powder during any fabrication route, and these fall into two groups—inorganic constitutents, and organic binders and lubricants. Conventional blending or mixing equipment can be classified into impellor types and tumble types;[76 77] the use of vibrational blending has also been studied.[78] Clearly the main purpose of blending is to ensure a uniform distribution of the additives, and therefore there is considerable interest in sampling and measuring techniques to assess the degree of mixing.[79,80]

The most common inorganic additives are those required to modify the sintering behaviour of the original ceramic powder. The roles of these additives can be classified into those that enhance densification and hence lower the sintering temperature, e.g. liquid phases, or impurities to enhance diffusion, and those additives that restrict grain growth, e.g. MgO in Al_2O_3.[81]

The addition of organic compounds to ceramic powders to assist the cold forming stage of fabrication is generally applied on an empirical basis specific to a given combination of organic and ceramic powder. Organic additives can be classified into lubricants that provide lubrication between particles and compacting die walls, and binders that improve adhesion between powder particles. Bruch[82] has studied the influence of organic lubricants on the green density of Al_2O_3 powders, and reported that coating the die wall with the lubricant gave improvements in green density comparable with dispersing the lubricant through the powder. Hart and Hudson[69] have reported a high green density for Al_2O_3 compacts formed by cold pressing; they obtained green densities of 75% theoretical density by the addition of only 3 w/o stearic acid. In addition to improving green density and hence reducing firing shrinkage, lubricants can reduce the pressures required for ejection from the compaction die and considerably lengthen die-life.

The role of binders on the strength of compacted ceramic bodies has been studied by Bruch,[82] who warned that improvements in crushing strength did not necessarily mean increased handling strength, and by Claussen and Jahn[82] who emphasized that certain organics can act as both binders and lubricants.

The common binders and lubricants can be added as a fine powder, or as a liquid or dissolved in a solution. Clearly the method of addition will determine the method of blending and also the subsequent powder treatment, such as granulation. Other additives that are necessary for certain forming stages of fabrication include plasticizers for slip-casting slurries,[84] or for extruding,[84] and also thermoplastic/thermosetting resins for warm-forming techniques such as injection moulding etc.[85] In the last case the thermoplastic binders can be added by warm blending using plastics technology.

Granulation and Screening

Two important features of powders are their flow properties, and their packing efficiency, and it is therefore common in the ceramics industry to enhance the flow properties and packing efficiency of fine powders by deliberate granulation into larger particles or granules prior to cold forming. The different techniques for granulation of a fine powder have been recently reviewed,[86] and include fluid-bed drying,[76] spray-drying[87] and pressing followed by crushing and sieving.[83] The influence of powder characteristics on the granulation process has been studied.[88] In most granulation techniques a binder is employed and therefore much of the previous discussion on binders is relevant to granulation.

The flow properties and packing efficiency of a granulated powder are determined by a number of powder characteristics including granule size and shape, and intergranular cohesion and friction.[89] The strength of the powder granules is also important; the granules must be strong enough to withstand handling, yet, if the granule structure is not broken under the compacting pressure during cold forming, then non-uniform microstructure will result.[82] The strength of granules has been studied by a number of workers.[82,90,91] Since the flow properties and packing efficiency of granules is markedly influenced by granule size distribution, the classification of granule sizes is an important step in ceramic processing. The various techniques, including screening through sieves and air classification, have been reviewed.[86]

SUMMARY

The stages of ceramic powder processing prior to cold forming and sintering, or hot pressing, have been reviewed. The emphasis of the review has been on the influence of the various methods of powder preparation and powder treatment upon the powder characteristics, since these characteristics will have a marked influence on all subsequent stages in the fabrication route.

Powders that have been prepared by calcination of chemically precipitated metal salts or hydroxides, require extensive treatment to produce the required powder characteristics for subsequent fabrication. New powder preparation methods are being developed which produce nearly spherical powder from solution by gel preparation,[92] spray drying,[61] freeze drying,[62] spray roasting,[63] sol gel calcination.[64,93] Powders have also been produced from the vapour phase either by vapour phase reaction[94–96] or by evaporation and condensation.[60,93,97,98] The advantage of these recent methods is that not only is the crystallite size small and therefore the powders very reactive but also as the powder characteristics are well controlled no other preparation is needed.

5

Powder Preparation of Hard Metals

The process technology of the cemented carbide industry differs considerably from that of most conventional powder metallurgy for two reasons: particle size and porosity.

PARTICLE SIZE

The powders are much finer than those used in most other processes. The particle size range typically describes grains as fine as 0·2–1·5 μm, and powders coarser than 8 μm are not normally used. This fine grain size results in powders which will not flow, have very poor pressing characteristics, are very difficult to mix homogeneously and are extremely sensitive to atmospheric conditions because of the very high surface area/weight ratio, commonly of the order of 5000 $cm^2\ g^{-1}$.

POROSITY

Porosity, which is the common characteristic of most powder metal products is completely unacceptable in many hard metal products, and in others must be kept very low.

First, porosity reduces the effective strength of hard metal by a considerable amount.

Secondly, porosity outcropping on to a surface gives holes which will spoil the appearance of the component being produced and give accelerated local wear. Porosity can result from many causes, but undoubtedly the two most important are pressing defects and impurities. Pressing defects are closely associated with defects in powder preparation. Impurities are probably the most serious source of trouble. The most dangerous would appear to be 'earthy' types of dust and those originating from structural materials such as cements, sand, brick, plaster, etc. Tiny particles of such materials react with the carbide at sintering temperatures with the production of very significant volumes of gas. The only way to avoid such problems is careful design of clean handling and processing conditions. This is especially important because holes as small as 10 μm, are commonly unacceptable.

These requirements result in the application of ball-milling and granulating in a way unique to this industry. The finished sintered materials consist of various mixtures of WC, TiC, TaC, NbC and Co. These are all produced as fairly fine powders, generally by simple chemical procedures. Simple blending of these constituents, even if the particle size is correct, will not give pressed compacts which will sinter without excessive porosity. In order to achieve the desired quality a very intensive wet ball-milling operation has been found to be essential.

Normally the operation is carried out in simple horizontally rotating ball mills, generally of stainless steel, but sometimes carbide lined. A charge of carbide balls some 2–3 times the weight of powder to be milled and a volume of liquid sufficient to convert the charge to a fairly thick slurry is normally used, and ball-milling times of 2–4 days are common.

Under conditions of ball-milling the combination of new surfaces and instantaneous high energy conditions can result in unfamiliar chemical reaction. Although no liquid is known that gives faster milling results than water there has been a general move towards inert hydrocarbon liquids to minimize the oxidation of the carbide and cobalt. Little has been done to explain the effect of different milling liquids on the speed of milling hard metal. Water is faster than the hydrocarbons, both in the rate of comminution of the coarser carbides and in the elimination of porosity after sintering. Comminution is usually associated with carbide of 2 μm grain size and above, and some adjustment of charge particle size is required to compensate for this. This technology has shown little change since the early days of the industry. Changes have been mainly aimed at increasing output by using larger mills or reducing milling time by the use of vibratory, or more recently, attritor type mills.

Whereas granulation may be used by choice in the ceramic industry, its use for hard metal pressing and sintering is essential. The as-produced powder is far too fine to flow naturally yet economic production of compacts in quantity on automatic tableting presses demands free-flowing properties. This is usually achieved by the addition of paraffin wax in a solvent to the powder, pressing the waxed powders, crushing and subsequently separating the particle sizes by sieving to give a suitable range for pressing. Considerable interest is being shown in methods of producing granules without pressure, and the industry is moving towards various vibratory and spray drying processes for the manufacture of granules. Spray drying was at first of little interest to the hard metal industry because the hollow shape of the granules resulted in serious porosity troubles, whilst the essentially larger scale of the operation was not readily accepted by the industry. However, the hollow granule problem has been largely overcome and the increasing scale of the industry has made the process very acceptable to the larger manufacturers.

6

Process Technology

SINTERING OF CERAMICS

Introduction

The process which develops the particular physical, chemical, electrical or mechanical properties which are unique in ceramic materials through application of heat is usually referred to as firing although the term sintering, originally used in the powder metallurgy industry, is increasingly used instead. During this process a number of physical, chemical and microstructural changes take place which usually lead to the creation of dense, strong, hard products. The driving force for these changes is surface free energy and this may act through the agency of diffusion through solids or migration in glassy liquids. Traditional ceramics, such as structural clay ware, fire brick and white wares, are almost universally densified by sintering in the presence of the liquid silicate phase which forms at high temperatures and serves as a structural bond during subsequent application. Solid state sintering has assumed importance with the development of newer materials such as pure oxides Al_2O_3, UO_2, MgO and mixed oxides such as ferrites and titanates.

Forming Processes

The main forming processes used in the ceramics industries are slip casting, extrusion, jiggering and pressing.

The Removal of Water and Binders—Drying

The operation immediately preceding firing and usually following directly after forming is a vital one which eliminates the liquid or plasticizer additions that have given the ceramic materials the properties needed for forming. In the initial stages of drying, where a continuous water film is present throughout the ware, the evaporation rate is that of water at a free surface. When this continuous film has disappeared the rate of drying decreases with decrease in water content as the transport of water from the centre of the body becomes more difficult. Shrinkage accompanies the first stage of drying and care is needed to prevent damage.

The rate of drying can be controlled by the amount, the humidity and the temperature of the air circulating over the ware. Infrared drying is used for some ware and recently direct firing with gas burners has been shown to give fast drying cycles.

The removal of binders is the equivalent operation associated with the dry press process. Whilst it can be incorporated into the heat-up stage of the main sintering furnace, this may be undesirable if the time for binder removal is long, since this may require excessive length in a tunnel kiln. Many binders are cellulose, wax or starch type products and can be conveniently removed by oxidation in air at a low temperature. The rate of removal needs careful adjustment as does the rate of drying, as excessive rates can cause disruption of the ware. When the bodies cannot be fired or heated in air the binders are often not completely converted to simple gaseous products and some provision for condensing tar-like products is necessary in the binder removal furnace.

Kilns and Furnaces

A description of the various kinds of kiln and furnace used for firing ceramic ware is to be found in Chapter 12.

Setting

This is the placing of the ware in the kiln prior to firing and the term also refers to the ware-stack itself. Bricks are usually placed in open checker-work arrangements such that the hot gases can get all around and through them. Care must be taken to ensure that the stress produced on the lower level by the brick load is not so high that it deforms at higher temperatures.

Saggars were frequently used in the Pottery Industry for supporting, containing and protecting the ware during firing. They consist of fireclay boxes, previously fired to a temperature higher than the firing temperature in the kiln. The ware may be set on sand or on edge supports so that it can shrink with the minimum of friction.

Open settings have tended to replace saggars since more ware can be fired and heat is not wasted in repeatedly firing the saggars. In those cases where open settings are used and the ware still requires protection, muffle kilns or electric kilns must be used. Open settings are arrangements of ware on refractory slabs and posts.

Firing

During firing, the ceramic ware is heated to a temperature between 700 and 2000°C over a cycle determined by the composition and the properties required in the product. A number of processes take place during this operation, some of them simultaneously, and although it is usually only necessary to provide close control over the maximum firing temperature it is often necessary to control the rate of heating and cooling and sometimes to provide a complete arrest during heating. Particular attention should be paid to the rate of temperature change when the following physical or chemical changes are occurring.

Removal of water

It might be thought that firing is always preceded by a drying process in which water is almost completely removed, but this is often not the case, and at the start of firing the ware usually contains several per cent and sometimes over 10% of water. It has been shown[109] that one type of unfired brick containing 2–2½% of water was likely to crack when introduced into a kiln which was being held at a temperature of 300°C.

Removal of binders and organic media

Binders usually decompose during the early stages of firing and provision is sometimes made to prevent the condensation of a tar-like product which, for example in sanitary fireclay kilns, may drip on to the ware and cause glaze faults.

Organic media used in on-glaze lithographic transfers must sometimes be decomposed and burnt slowly to avoid frizzling. Slow rates of heating in the 200–500°C range have been recommended.[110]

Dehydroxylation

Clay minerals usually decompose with evolution of steam between 500 and 600°C. It has been shown[111] that the loss of strength that occurs at this stage can result in the cracking of completely dried articles.

Oxidation

The oxidation of carbon and sulphur compounds must be completed before densification is far advanced if black cores and bloating are to be avoided. This is achieved most rapidly by arresting the firing for a period at a temperature that varies with the type of body and the making method employed but which is usually about 900°C.[112,113]

Decomposition

The decomposition of carbonates and sulphates may produce bloating in vitrified bodies.[114]

Phase transformations

Silica is an important constituent of most ceramic bodies and it exists in many crystalline forms.[115,116] The conversion of one form to another, which is accompanied by large volume changes, occurs during heating and cooling and the rate of temperature change may have to be reduced.

Solid phase sintering

As pressed, a solid compact contains 30–50% porosity and in order to achieve the desired levels of strength and other properties it is usually necessary to reduce this to less than 5–10%. By the use of fine powders and temperatures which are high, but well below the melting point, it is possible to achieve substantially complete pore removal by diffusion or plastic flow processes.

An illustration of the complex reactions and transformations involved in firing ceramics is given in the following description of the effect of heat on Kaolinite [$(OH)_4Al_2Si_2O_5$] which is a major constituent of many ceramic bodies. Kaolinite crystals are largely unchanged, but lose a little weight on heating to 400°C. At 450°C they break down at constant temperature forming a structureless mass which bears some resemblance to the original crystals. Water is eliminated here with a considerable loss of weight and intake of heat but the atomic-scale structure is little changed. At 980°C the amorphous mass crystallizes suddenly into mullite with a large evolution of heat but no change in weight. As the temperature rises still higher the mullite crystals grow and the glassy bond begins to operate to hold the grains together producing shrinkage. At about 1200°C, cristobalite, a high-temperature form of silica, crystallizes from the glassy bond.

The Activated Sintering of Ceramics

The term 'activated sintering' is used here to mean a sintering process which has been deliberately designed to improve the sintered density, or the densification rate, of the material concerned by means other than the well known and fundamental ones of the application of pressure or the increase of sintering temperature or time.

The principal ways in which such activation can be brought about are by the increase of surface and volume diffusion rates, by the removal of some barrier to diffusion existing at the particle surface; and by the prevention of rapid grain growth from trapping pores within grains.

The increase of bulk diffusion rates can be brought about by the use of additives which introduce either distortions or vacancies in the crystal lattice, by the use of sintering atmospheres to control the stoichiometry of the material and by the physical distortion of the lattice prior to sintering by mechanical deformation. Surface diffusion —which can be important in the early interparticle neck formation stage, but which cannot contribute to densification directly—can be enhanced by sintering atmosphere, additives, and by the formation of very thin adsorbed or chemisorbed layers on the particle surface.

The removal of surface barriers to diffusion generally involves the reduction of an oxide layer by molecular or atomic hydrogen in the sintering atmosphere or the reaction of the layer with a solid or liquid additive.

The rate of grain boundary movement in a sintering body is reduced if the boundary is associated with a concentration gradient of a solute in the matrix. This is because the boundary movement requires not only the normal transposition of matrix atoms between the differently oriented lattice positions on either side of the boundary, but also the diffusion of the solute atoms over much larger distances. The importance of retarding grain boundary movement, and hence discontinuous grain growth, in order to attain sintered densities of over about 95% of theoretical, has been pointed out by Coble and Burke.[19]

The use of activated sintering in the ceramic industry is not widespread, but has achieved importance in those areas where the fabrication of materials having a very high melting point or to a very high density is necessary. The following Table lists the more common examples.

TABLE 3.

Substance	*Activation method*	*Reason for use*
Al_2O_3	MgO grain growth inhibitor	To obtain 100% density
MgO	LiF	To obtain 100% density
UO_2	Control of atmospheric O_2 potential	To reduce sintering temperature and increase density
UO_2	Addition of TiO_2	To reduce sintering temperature and increase density
	Addition of Nb_2O_3	To reduce sintering temperature and increase density
PuO_2	Fe	To reduce sintering temperature
UC	Addition of Ni	To reduce sintering temperature
ZnO	Addition of NiO or CoO	To increase density

7

Reaction Sintering

INTRODUCTION

The majority of present-day ceramics are made by taking the finely powdered ceramic material, shaping it and sintering the powder together by heat treatment.

In most cases, this heat treatment gives an adequately strong article and can, if needed, give a fully dense one as well, though often this is at extra expense.

However, in some areas, this simple technique is unsatisfactory, as the powder may not sinter together, the sintering temperature or atmosphere requirements may be impractical or uneconomic, or the basic material may not be readily available or may decompose under normal sintering conditions. In some of these cases, reaction sintering may offer the possibility of making sound ceramics which are impossible or difficult to make by other methods—often of high density as well.

GENERAL

The term 'reaction sintering' is not very well defined and could be applied to three different sintering processes.

In the simplest case, it can be applied to the technique in which a powder that decomposes on heating to form the required final compound is substituted for the normal powder in a conventional sintering operation. Though, for completeness, this process is considered briefly later, it is doubtful if it can truly be considered reaction sintering, at least in the author's opinion.

In the other two cases, the two (or more) components of the required ceramic compound react together during the sintering operation. In the first of these cases, the two components, in powder form, are mechanically mixed, shaped and reaction-sintered together. This process is suited to compounds such as carbides, borides, silicides, aluminides, etc., in which both constituents are solids at room temperature and of fairly high melting point.

In the second case, applicable where one of the constituents is gaseous at room temperature (oxides, nitrides) or of low melting point relative to the other constituent (e.g. phosphides, sulphides and some silicides or aluminides), the more refractory constituent in powder form is shaped and reacted with the other constituent in gaseous or liquid form. Because of the need for the gaseous or liquid constituent to react with the other constituent, this latter constituent must be porous, both to allow entry of the reactant and to allow for the extra volume (if any) of the reaction product.

If there is an increase of volume, this does give, at least in theory, the possibility of making dense materials without sintering shrinkage (though in practice this may be limited by pore blocking and diffusion rates, especially in larger pieces).

Shaping methods for both of these cases follow normal ceramic processes such as pressing, slip casting, extrusion and plastic moulding,[117–120] while flame spraying has been used to make the silicon preform for silicon nitride.[120]

SOME METHODS OF REACTION SINTERING

Using Decomposable Raw Materials

Traditional particles such as pottery, made from clay which decomposes at about 500–600°C, losing water and changing in composition, could be considered to fall within this category.

However, the process is more usually considered to apply to purer materials such as magnesia, which can be made from magnesium carbonate or hydroxide.

Essentially, this technique gives a very active powder with enhanced sinterability, and in other respects the factors, kinetics, etc., are analogous to normal sintering operations.

In many cases, because of the inevitable high shrinkage found in this method, which tends to give flawed pieces with poor dimensional control, the decomposable powder is pressed during the sintering operation to obtain a sound article.

This technique has been applied recently to magnesia,[121,122] alumina,[121,123] thoria[121] and clay,[123,124] though the idea of hot-pressing clay was proposed as early as 1939.

Using All the Constituents in Premixed, Solid Form

Under normal slow sintering conditions it seems impossible to use premixed reacting powders because the product is likely to be inhomogeneous and have rather low 'green' (i.e. unsintered) density. Some results on silicon carbide made from silicon and carbon powders have been reported and confirm this observation.[126]

However, if the premixed powders are heated up rapidly the heat of reaction of the two components can be used to assist sintering. Rapid heating is necessary to initiate the exothermic reaction, which otherwise takes place very slowly with no useful increase of temperature.

This rapid heating technique has been used in the case of uranium carbide, UC.[127,128] In this example, the original uranium and graphite powders can be compacted to a fairly high green density and there is only a small expansion on reaction sintering so that the final product, though not fully dense, is still about 90% theoretical density.

In many other cases, however, where there may be quite large volume changes or the green density is low, simultaneous pressing and reaction sintering gives better results and may also eliminate the tendency to flaws. Fully dense ZrB_2, $MoSi_2$, NiAl and composites containing these materials as the matrix have been prepared successfully in this way from (Zr + B), (Mo + Si) or (Ni + Al) powders.[129]

Using One Constituent in Solid Form and the Other in Liquid or Gaseous Form

This technique is particularly useful for making high density binary compounds that sinter poorly under normal conditions, e.g. silicon nitride, silicon carbide, but is not, of course, limited to such materials.

Alumina has been made from a porous aluminium shape reacted with oxygen gas and other oxides have been experimented with, but in general oxides do not appear to be very well suited to this technique.[130] Nitrides give better results, however, and titanium nitride,[130] silicon nitride,[131,132] chromium nitride, Cr_2N[130] and aluminium nitride[130,133] have all been made successfully using the porous metal shape reacted with nitrogen. In the case of AlN the low melting point of aluminium restricts the nitriding temperature until a skeleton of AlN is formed which will hold together and prevent liquid Al running out. However, as the nitriding rate of Al is slow below its melting point temperatures as near as possible to this are used, but this tends to give faulty specimens with flaws where the Al has melted out. This problem has been overcome by using a mixture of preformed AlN and Al powder instead of Al powder only.[129]

Dense silicon carbide has also been made by treating a powdered graphite–silicon carbide mixture with silicon, either as a liquid or a vapour,[134] while boron phosphide, BP, has been made by reacting phosphorus gas with a powdered boron shape.[126]

Silicon nitride has also been made from silica and carbon powders mixed in the ratio $SiO_2:2C$, shaped and reacted with nitrogen gas.[135] This can be considered a combination of reduction and nitriding but the low density of the product due to the large weight loss of CO makes it rather impractical for most uses.

THEORETICAL CONSIDERATIONS OF REACTION SINTERING

Using Decomposable Powders

As this is essentially the same as normal sintering or hot pressing, it is not here considered.

Using All the Constituents in Premixed, Solid Form

Stringer and Williams[129] have described the requirements for reaction sintering in this case. The first is a substantial heat of formation, of the order of 10 kcal per g atom of what they term the 'active' component, i.e. the more rapidly diffusing element. This is necessary to give a peak temperature sufficient to give good reaction.

Rapid heating and a low surface area in relation to volume are also necessary so that the useful exotherm is not dissipated before the runaway condition is reached. Heating times of 2–10 min are typical.

A liquid phase during the reaction can be useful in promoting complete reaction, e.g. by solution/precipitation or, as in the case of UC, by intergranular corrosion which breaks up the reaction product.[128] The exotherm initiating temperature should therefore not be too much below the liquidus temperature, and phase equilibrium studies and wetting behaviour knowledge are helpful here.

Complete reaction is also helped by fine particle size though this may necessitate still faster heating to prevent loss of exotherm.

In order to allow densification to occur by shrinkage (whether or not pressure is applied) the reaction product should not be too strongly deformation resistant. As the exotherm happens very quickly, if pressure is applied, the press must be very fast-acting and free-moving otherwise maximum density may not be obtained.

Using One Constituent in Solid Form and the Other in Liquid or Gaseous Form

The requirements for this process are discussed by Gooding and Parratt[130] and Popper.[126]

The most obvious requirement is that complete reaction should not be able to occur inside the compact—in other words, the outside of the compact should not become blocked by reaction products (or unchanged material transported by evaporation, etc.) before the inside has completely reacted.

Gooding and Parratt[130] identify two types of reaction:

1. The solid retains its particle shape and the gas (or liquid) converts it to a compound particle.
2. Very little of the original solid remains as it evaporates and reacts in the pores with the gas (or liquid) leaving a void where the solid originally was.

In general, both types of reaction occur together.

In the first reaction mechanism (unless the reaction product sinters or reacts together) the initial porous solid compact will need to be sintered in order that a bonded reaction product is obtained. The initial porous compact will also expand in proportion to the volume change solid→compound, though if this is excessive it may cause disruption.

In the second reaction mechanism, there will be no gross volume change and even a loose solid powder compact will become bonded. However, in both cases, porous products will be obtained.

A balance of the two mechanisms is probably desirable and could ideally give full density. In fact if there is no external dimensional change it can readily be seen that full density is obtained if $(1-P)VR = 1$, where P = fractional porosity of solid before reaction sintering and VR = true volume ratio of original solid to final product (after allowing for porosity).

Some volume ratios are given in Table 4.

TABLE 4.

Solid	*Volume ratio—compound: solid* Oxide	Nitride	Silicide
Aluminium	1·28	1·26	—
Chromium	2·03	1·07	$CrSi_2$ = 2·99
Magnesium	0·81	—	—
Silicon	2·2–2·3	1·26	—
Titanium	1·76–1·95	1·07	$TiSi_2$ = 2·37
Zinc	1·59	—	—
Carbon	—	—	2·2–2·35

Silicon nitride appears to give a balance of the two mechanisms whereas aluminium nitride tends to form mainly by the second mechanism. Thus, dense AlN is better made by nitriding a mixed Al–AlN powder. Titanium nitride, on the other hand appears to form mainly by the first mechanism and the porous titanium preform must therefore be sintered. As the volume expansion is not great, sound articles can readily be made.

Aluminium oxide forms normally by the first mechanism, but the rate of oxidation of porous Al compacts is extremely slow, using air or oxygen. Alumina powder mixed with the Al gives an improvement but a low oxygen potential gas, by allowing the Al to vaporize before oxidation, will tend to favour the second mechanism and could perhaps give a high density product.

Carbon silicide (silicon carbide) has a high volume increase on forming though the volume increase of porous carbon shapes on siliciding is not too large suggesting both mechanisms operate. As, however, carbon seems unlikely to vaporize it is more likely the effective reaction in the pores is the result of liquid Si corroding away the reaction product by intergranular attack and depositing this SiC in the pores.

The C + Si compact expansion, though small, is, however, often sufficient to crack the brittle compact and better results are obtained by mixing prereacted silicon carbide with the porous carbon preform. In this way almost fully dense silicon carbide has been manufactured.

KINETICS OF REACTION SINTERING

These tend to be specific for each system and have been discussed mainly in relation to the two most economically attractive materials, silicon carbide and silicon nitride.

In the case of silicon carbide[136] the rate of penetration of the liquid silicon into the porous carbon appears to be the deciding factor of the rate of sintering and this is governed by pore size and the surface tension, contact angle and viscosity of the liquid silicon. This gives a dependence of height (h) silicided against time (t) such that $h^2 = kt$. Owing to the exothermic nature of the reaction, very high local temperatures may exist, particularly in the initial reaction zone, and this can lead to cracked areas.

Much more work has been carried out on the reaction kinetics of the silicon–nitrogen reaction[137–142] though this is complicated by the presence of two nitride phases, $\alpha + \beta$, the former oxygen-stabilized. The reaction, like Si + C, is also exothermic and care is needed to avoid melting the silicon in the early stages of nitriding. The effects of pore structure, silicon particle size, size of piece undergoing nitriding, reaction temperature and time, impurities and catalysts are all important and have been investigated by the authors quoted in Refs 137–142. The observed parabolic rate has been considered to be controlled by the permeation of nitrogen through the porous matrix, or alternatively by the silicon–nitrogen reaction rate. The latter seems the more likely, but the former will eventually prevail as the size of article being nitrided becomes greater.

ECONOMIC IMPORTANCE

There are only two materials formed by true reaction sintering of economic importance: silicon nitride and silicon carbide.

Neither of these has a very large market, however, and tentative estimates of world sales are of the order of £100 000–200 000 p.a. for both materials in 1971. The potential could, however, be much larger, possibly in the £1–10 million p.a. range.

Some Outstanding Problems and Research Requirements

The following problems exist for reaction-formed silicon nitride:

1. More knowledge of the nitridation process as carried out in bulk is needed to establish optimum nitriding schedules.
2. Improved furnace design is needed which will reduce the cost of the high-temperature nitridation stage.
3. Methods of making reaction-formed silicon nitride fully dense are needed, e.g. vapour cracking more Si_3N_4 in the pores.
4. A better understanding of the links between fabrication methods, microstructure including α–β ratios and properties is needed.

The following problems exist for reaction-formed silicon carbide:

1. More knowledge of the Si–C reaction is needed to enable fully dense materials to be made without excess of free silicon or free carbon.
2. The technology of manufacturing thick fully dense sections, with optimum properties, needs further development.
4. Better understanding of the correlation between manufacturing methods, microstructure and properties is needed.

There is a need to find more applications for reaction-formed silicon carbide and silicon nitride in order to break the vicious circle of low usage causing high costs due to low production, as both materials are potentially fairly cheap (although they will never be as cheap as conventional ceramics, such as alumina, except where machining costs are very high).

8

Fibre-reinforced Materials

INTRODUCTION

Powder metallurgical techniques can be used in the production of fibre-reinforced metals and particular emphasis has been placed on methods capable of actually producing components. The discussion is mainly limited to reinforcement of metals with available advanced continuous fibres such as silica, boron, silicon carbide, carbon and beryllium, but some mention will also be made of reinforcement with strong ceramic whiskers such as alumina, silicon nitride and silicon carbide. Earlier work is described by Cratchley.[143]

The requirements of a fabrication process for fibre-reinforced metals are almost identical with those for reinforced plastics. For instance, the processes must (a) not damage fibres, (b) allow accurate positioning of the fibres, (c) allow control on the fibre concentration and distribution, (d) permit the fabrication of a variety of shapes and sizes of component, and (e) provide a matrix with the desired properties.

The economic factors must also be favourable and this depends on the application of the composite. In general the aim of reinforcing metals is to extend the very attractive specific properties of reinforced plastics to higher temperatures, although there are a number of other very important advantages such as the decreased anisotropy of their strength and elasticity. For aerospace type of application the higher material and fabrication costs may be relatively easily justified provided the predicted properties can be obtained.

CONTINUOUS FIBRE COMPOSITES

The two main techniques used for fabricating plastic matrix composites with continuous fibres are filament winding and warp sheet lamination. The first method consists of winding the fibres on to a rotating mandrel; resin is applied either during winding (wet process) or as a semi-cured coating on the fibres (pre-preg). The second process consists of first preparing a sheet of semi-cured resin containing fibres (warp sheet) and then laminating these sheets by hot pressing (similar to plywood), to form the final component. Thus simple solids of revolution such as pressure vessels can be made by the first process, and plate-like shapes such as aerofoils can be made by the second. A third process used to a much lesser extent with continuous fibres is injection moulding. In this case the fibres are pre-aligned in a mould usually under reduced pressure and the resin is injected under pressure.

All of these processes are applicable at least in principle to metal matrix composites. The major limitation is the avoidance of fibre damage which can occur either by chemical interaction of the fibre with the metal matrix at elevated temperature, causing weakening, or mechanical fragmentation at high pressures. To a large extent this rules out the injection moulding method because the majority of the fibres are weakened by contact with molten metals. Attempts have been made to develop a filament winding technique applying the metal either by plasma spraying[144] or by electrodeposition.[145–147] Electrodeposition is in some respects an ideal method since many metals can be applied at ambient temperatures and pressures as a continuous matrix.

The warp sheet method has however been by far the most successful and this relies on one or more stages of diffusion or liquid phase bonding. The warp sheet may be produced directly either by wire or foil bonding, electrodeposition or plasma spraying or indirectly by first pre-coating the fibres with metal and then using an organic bonder to get these into the form of a warp sheet. The composite component is produced from the warp sheet by a subsequent laminating operation which usually involves either direct hot pressing or braze bonding under pressure.

PARTICULAR SYSTEMS

To be of any practical interest the fibres and metal matrix must be chemically compatible, the metal must protect the fibres from oxidation and have the desired physical, mechanical and chemical properties. When the various possible systems are compared on this basis it would seem that aluminium, and to a more limited extent titanium, are the only matrices of major interest. This of course limits the temperature capability of metal matrix composites and suggests the urgent need for further fibre development—alumina fibres may be one possibility. It is convenient to consider the various metal matrix composites under the headings of the individual fibres. Details of these fibres are given in Table 5.

Silica Fibre-reinforcement

Although silica fibres have a low density and can have very high strength (Table 5) they have a modulus of only 72 MN m^{-2} which makes them of very limited value for metal matrix reinforcement. A technique of producing silica fibre reinforced aluminium was developed at Rolls-Royce Ltd in about 1962 before other advanced fibres were readily available. In this process silica fibres ($\sim$50 μm diameter) coated with aluminium (about 50% by volume) were prepared by passing the fibre immediately after drawing, through a specially designed furnace containing molten aluminium.[148] Because the highly surface-sensitive fibre was immediately protected by the aluminium coating very high strength was retained (although not as high as could be obtained with a polymer coating).

Composites were produced from the coated fibres by hot pressing and the effect of the hot pressing variables, pressure, time, temperature and environment, were investigated in some detail. The main difficulty was to obtain a sufficient bond between the

TABLE 5.

Fibre	*Form*	*Manufacturing process*	*Specific gravity*	*Modulus* $GN\ m^{-2}$	*Strength* $GN\ m^{-2}$	*Comments*
Silica (SiO_2)	Monofilament, about 50 μm diameter	Viscous drawing of silica rod	2·5	72	7·0	Highly surface-sensitive; must be immediately coated to retain strength; limited by low modulus and by viscous creep at elevated temperature (800°C).
Beryllium (Be)	Monofilament, about 125 μm diameter	Powder metallurgy and wire drawing	1·85	—	1·1	Fairly low yield stress; creeps at 300°C; very expensive.
Boron, Silicon Carbide, (B), (SiC)	Monofilament, about 125–150 μm diameter	Chemical vapour disposition to a tungsten wire	2·3 3·2	350 ~450	2·8 2·1	In many respects ideal for aluminium reinforcement but expensive.
Carbon (C)	Multifilament, about 8 μm diameter	Pyrolysis of a polymer precursor heat-treated to ~1000°C or ~2500°C	1·8	240 350	2·8 2·1	Very fine multifilament form makes matrix fabrication very difficult; potentially relatively cheap.

aluminium coatings without causing chemical or physical fibre damage. In the first fabrication experiments[149] carefully aligned coated fibres were hot pressed in a tool steel mould in air and the relationship between pressing pressure and temperature with strength is shown in Fig. 5(a), and a simple self-explanatory model is shown in Fig. 5(b). The difficulty of bonding aluminium is due to the oxide film and the pressure and temperature conditions must at least be sufficient to cause partial breakdown of this film. However, at temperatures above 550°C the fibres are severely degraded by reaction with aluminium to produce silicon and alumina and the fibres are very easily mechanically damaged by excessive pressures. Thus a considerable effort was put into methods of minimizing the temperature and pressure requirements. In subsequent experiments hot pressing was carried out in a vacuum chamber (often with a low hydrogen pressure to compensate for poor vacuum) and a considerable reduction in temperature and pressure for a given matrix bond strength was found possible. This improvement is shown in Fig. 6(a) and (b) where the strength normal to the fibre direction was used as a measure of the bond strength. A considerable improvement in composite strength was obtained from the reduction in fibre damage (from 2·8 to 4·6 MN m^{-2}) although it was not found possible to remove entirely the boundary between the original aluminium coatings (see Fig. 7).

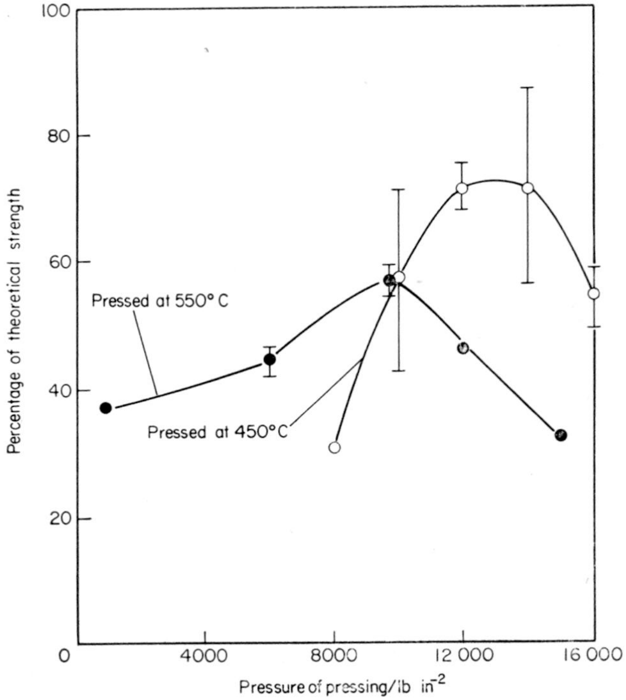

Fig. 5(a) Effect of temperature and pressure of pressing on ultimate tensile strength of composites.

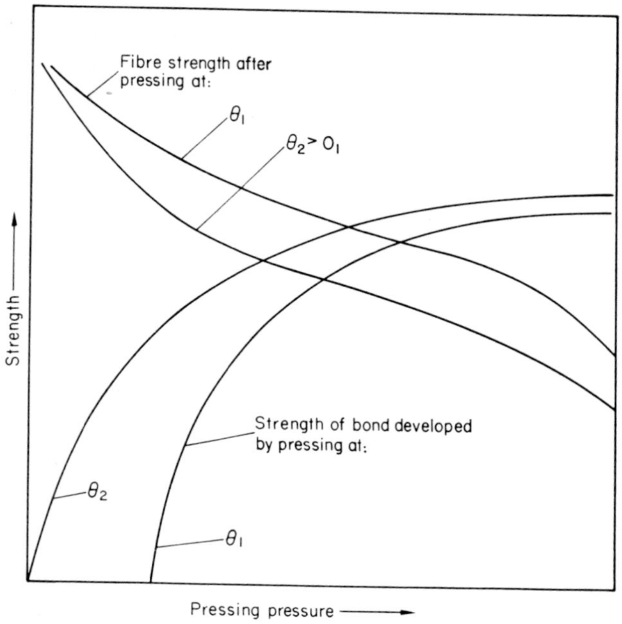

Fig. 5(b) Model proposed for the effect of temperature and pressure of pressing on the strength of composites.

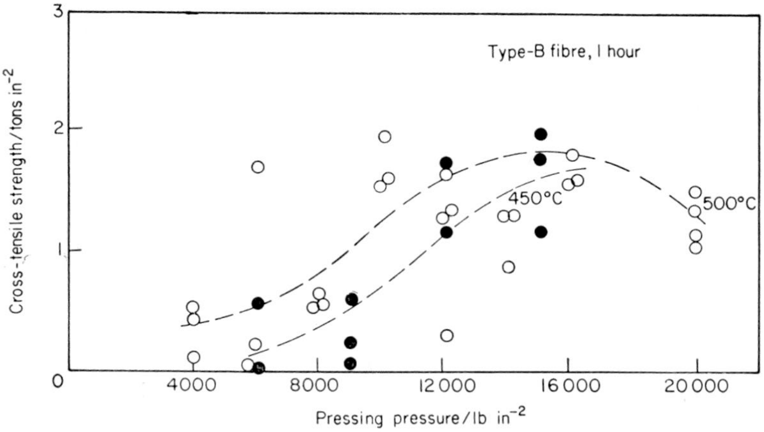

Fig. 6(a) Variation in composite cross-tensile strength with pressure and temperature (Type-B fibre). ○—500°C; ●—450°C.

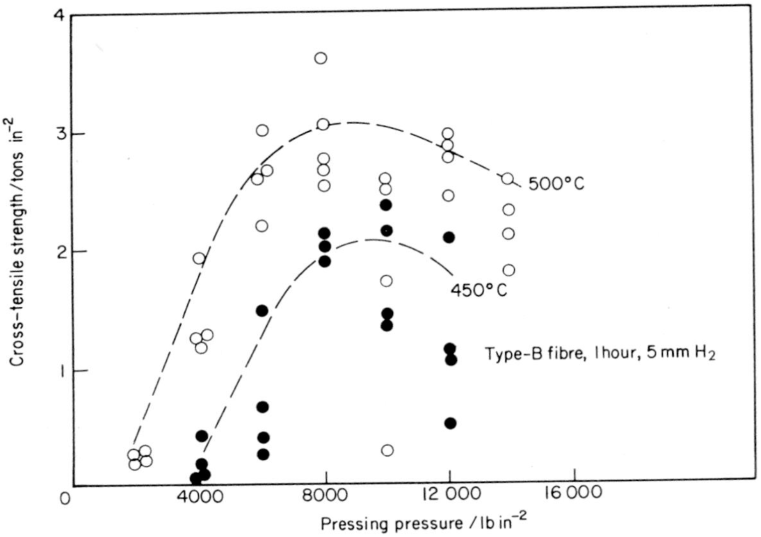

Fig. 6(b) Variation in composite cross-tensile strength with pressure and temperature (Type-B fibre, H_2). ○—500°C; ●—450°C.

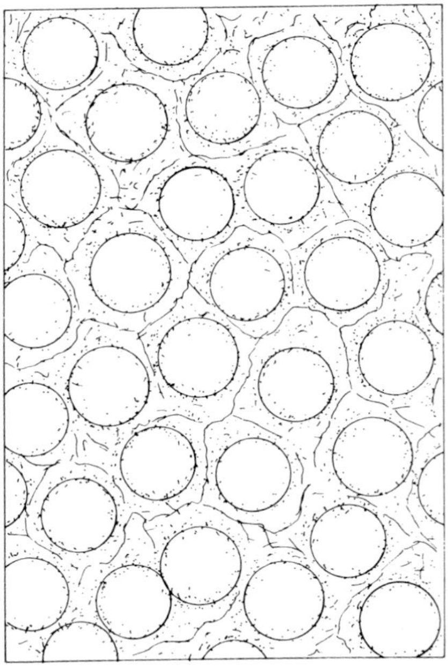

Fig. 7. Microstructure of SiO_2 fibres coated with Al pressed at 500°C with a pressure of approximately 6·9 MN m^{-2}.

Components were produced[150] by filament winding and hot pressing (usually these were in the form of narrow rings) and also from warp sheet. The warp sheet was produced by winding the fibres on to a drum and then bonding them together with a removable organic binder; the bonded sheet was then slit and removed from the drums for laminating. Component shapes such as aerofoil sections were constructed by cutting the warp sheet to the desired shape, stacking in a mould, heating to remove the binder, and then hot pressing. Where the warp sheet layers were at alternating angles to each other it was found that the fibre damage could be considerably reduced by using aluminium foil interleaves.

Beryllium Wire Reinforcement

Beryllium has a low density and high modulus and can be produced as a fairly strong wire ($\sim 125\ \mu m$ diameter). In addition, unlike all the other advanced fibres, it is relatively ductile although this is apparently not sufficient to overcome the problem of brittle fracture in components produced from the bulk material. The wire form is at present very expensive.

Beryllium wires have been used to reinforce aluminium and titanium using hot pressing as a means of composite fabrication. The fibres are not as prone to mechanical damage during pressing as brittle fibres, but fibre weakening due to annealing the formation of intermetallics by fibre–matrix interaction is a problem.

Composites have been prepared mainly by using metal powders mixed with aligned filaments[151] and by using metal foils.[152,153] The powder technique is limited by the difficulty of obtaining good fibre distribution. The foil-bonding technique consists of laying up wires (either by hand- or filament-winding) between layers of the metal and hot pressing usually in vacuum. The foils then extrude between the fibres and bond along the filament centre line; presumably the high deformation in the extruded metal aids oxide film break up. For subsequent laminations monolayers can be produced to form warp sheet.[113] Alternatively the composite can be pressed in one operation with the filaments stacked with the desired orientations. Typical hot pressing conditions are approximiately 4000 lb in^{-2} ($\simeq 28$ MN m^{-2}) at 500°C for beryllium fibres with 7002 aluminium alloy sheet and 13 000–15 000 lb in^{-2} ($\simeq 90$–104 MN m^{-2}) at 600°C for titanium alloy sheet Ti 6 Al-4v. It would appear that some reaction does occur during hot pressing presumably between fibre and metal but with ductile fibres the effect of this may not be as serious as with a brittle fibre; certainly with silica fibres, probably the most surface-sensitive of all the fibres, a catastrophic drop in strength occurred with submicroscopic quantities of reaction product.

Boron, Borsic and Silicon Carbide Filaments

Boron and silicon carbide can be produced in a very strong high modulus form by vapour phase reaction and deposition[154] on to fine tungsten wire ($\sim 8\ \mu m$ diameter). The filaments are usually produced with a fairly large diameter (125–150 μm) to overcome the high density of the tungsten core; this is in fact converted to tungsten boride. Both of these fibres can only be considered for application in an aluminium or titanium matrix because of the poor compatibility with the higher temperature metals such as nickel. Silicon carbide is more resistant to interaction with titanium and aluminium at high temperatures and also has a better oxidation resistance. It is, however, more expensive to produce and is usually not as strong as boron. Thus a silicon carbide coated boron filament (called Borsic) has been produced in large quantities particularly for metal matrix application. Boron and Borsic fibres are being produced in the U.S.A. in large quantities but are still fairly expensive.

Composites in an aluminium alloy matrix are produced from preformed warp sheet which may be produced by plasma spraying[155] or by foil bonding[156] as described previously. In both cases the boron filaments are either carefully hand layed up or filament wound to obtain good distribution. Often the fibres are held in position by means of a removable binder. The warp sheets are then either diffusion-bonded by hot pressing (typically 550°C at 5500 lb in^{-2} (38 MN m^{-2}) for one hour in vacuum) or brazed with an aluminium silicon alloy in vacuum under pressure. The plasma-sprayed matrix (typically 600°C at ~50 lb in^{-2}) is porous and requires deformation by hot pressing or infiltration by a liquid braze to convert it into a useful matrix; even so its properties are never as good as the foil-bonded material. Similar techniques have been used to produce joints between preformed composites.[157] In fact a warp sheet or Borsic fibres in a plasma-sprayed matrix (6061) aluminium alloy is commercially available from Hamilton Standard. This may be backed with an aluminium silicon brazing foil or an aluminium alloy foil for diffusion bonding.

The fabrication technology with titanium[158] is similar although only the foil- or sheet-bonding technique has been reported. However, one modification used is to place the filaments in pre-rolled grooves in the titanium sheet prior to hot pressing.[158] This considerably reduced the temperature, time and pressure requirements for bonding because large scale flow of the matrix is not required and thus allows the warp sheet to be formed by a high speed diffusion bonding process.

The effect of the hot pressing conditions on fibre damage in aluminium–boron and silicon carbide composites is described by Cunningham *et al.*[159] Besides outlining the temperature and pressure conditions required to minimize fibre breakage and further matrix interaction to form aluminium borides or silicide, it is also shown that fibre strength degradation occurs by oxidation unless pressing is carried out in an inert atmosphere. A more detailed investigation of the effect of temperature or intermetallic formation in the boron–titanium and silicon carbide–titanium systems is described by Metcalfe and Schmitz.[158] The compounds formed are either titanium diboride (TiB_2) or titanium disilicide ($TiSi_2$). It is suggested that these are not detrimental to the filament strength provided they do not exceed about 1000 Å in depth. In fact the high speed diffusion bonding process described earlier produces about 200 Å of compound at the fibre matrix interfaces.

More recently a boron filament with submicron nitride (by reaction with ammonia and hydrogen) coating has been produced which is reported to have substantially improved compatibility with metals and improved oxidation resistance.[160] This should simplify hot pressing and may also make some of the liquid metal fabrication techniques more attractive.

Carbon Fibre Reinforcement

Carbon fibres have a low density and can have a high strength and stiffness and are available with various combinations of modulus and strength (see Table 5, p. 35). They are produced by pyrolysis of a polymer precursor fibre and are in the form of multifilament tow, usually about 10^4 filaments, with an individual filament diameter of $\sim 8\ \mu m$. These fibres are very different from those discussed so far, which were monofilaments with diameters greater than 50 μm, and present some very difficult metal matrix fabrication problems. Although carbon fibres are reasonably compatible with nickel and other high-temperature non-carbide-forming metals,[161] these metals do not affect protection from oxidation, even at relatively low temperatures[162] ($\sim 550°C$). Strong carbide-formers such as titanium cause weakening of the fibres by chemical reaction even at relatively low temperatures.[163] Thus aluminium is the only matrix of any interest since this can protect the fibres from oxidation[164] and does not readily form carbide at temperatures below 600°C.

Various methods of producing aluminium carbon composites have been tried. So far the best mechanical properties have been obtained from composites produced by infiltration with aluminium–silicon.[165] However, this technique is of limited application for component fabrication because of the difficulty in controlling fibre orientation and in producing large thin-walled structures; in addition, the use of an aluminium–silicon alloy limits the high temperature application and reduces the corrosion resistance of the composite.

Methods of producing a carbon fibre–aluminium matrix warp sheet starting material for consolidation by hot pressing have been developed. To obtain the high penetration required to coat a multifilament tow, both winding, molecular forming techniques and electrodeposition[166] have been used. These are chemical vapour deposits.[167] With chemical vapour deposition 10^4 filament tows can be easily coated and subsequently consolidated by hot pressing. The penetrating power of electrodeposition is much more limited and only tows of 10^3 filaments can be processed, but the resultant material is a continuous matrix warp sheet.

The coated fibres (chemical vapour deposited) or warp sheet (electrodeposition) are consolidated by hot pressing in vacuum. Typical pressing conditions used are 10 000 lb in^{-2} (96 MN m^{-2}) at 550°C for 1 h or alternatively 500 lb in^{-2} (3·5 MN m^{-2}) at 600°C for 1 h.

The main difficulties encountered in hot pressing are mechanical fibre damage at high pressures (low temperatures) and the formation of aluminium carbide at high temperatures (low pressures). Fibre damage is unavoidable in multifilament tows in which near-perfect fibre alignment cannot be maintained because of the extreme pressures at cross-over points. The controlled formation of aluminium carbide is a useful means of improving the fibre matrix bond strength and does not result in fibre strength loss. However, too strong a matrix bond strength in this system has been found to result in poor mechanical properties in the composite, even though the fibres are unweakened. Of course, gross carbide formation at temperatures above 600°C results in severe fibre weakening.[168]

Discontinuous Fibre Reinforcement

Under certain conditions it is possible to grow a number of ceramic materials in the form of ultra-high strength single crystal whiskers. These are extremely fine discontinuous filaments, typically with lengths of a few hundred microns and diameters of 1 or 2 μm, which owe their strength to the absence of internal and external flaws.

To make use of the high strength obtainable from these filaments it is necessary to use them as a reinforcement in a suitable matrix. With discontinuous filaments the matrix must be capable of transmitting load into the filaments by a process of shear around the fibre ends.[169] The shorter the filament for a given diameter the greater the shear stress the matrix is required to provide to load the fibre fully. Thus for most whiskers which usually have low average aspect ratio (length:diameter) metals are a more suitable matrix choice than plastics.[170]

The composite fabrication problems are, however, formidable since the whiskers are produced as a woolly mass, usually with a large amount of non-whisker dust, and the whiskers themselves usually have a very wide range of aspect ratio. The first problem is to sort out the useable whisker lengths from the woolly mass and then devise some method of orientating them and finally to get them into the matrix with minimum fibre damage.

Whisker sorting is usually done by screening and washing to remove dust and to separate the matrix into various lengths.[170] A number of methods are used to obtain a high degree of orientation. For instance the whiskers may be mixed in a dope and extruded; the matrix may also be added to the dope in the form of a fine powder. Thus Schierding and Deex[171] mixed silicon carbide whiskers with aluminium powder in a dope made of cellulose acetate dissolved in methyl acetate and extruded this through a venturi into a solvent coagulating bath. Another method[172] is to mix the whiskers, in this case alumina or silicon carbide, with an aluminium–silicon alloy powder and filter in a magnetic field. For the alumina whisker, wire is made magnetic by applying a thin coating of nickel by chemical vapour disposition from nickel carbonyl.

Although various methods of consolidation have been used, extrusion and hot pressing are the methods used the most. Extrusion,[170] as may be expected, improves fibre orientation and indeed it is possible to use this as the only means of orienting the fibres. However, it does cause whisker fragmentation.

Again the main problem in hot pressing is to avoid mechanical or chemical fibre degradation. Fortunately the whiskers considered are reasonably stable in aluminium and aluminium alloys, which again is probably the only metal of technological interest for whisker reinforcement; magnesium is, however, another possibility.

The effect of hot pressing on whisker damage is considered in some detail by Schierding and Deex[171] for the silicon carbide–aluminium system described above. With a starting average aspect ratio of 124 at 50 volume percent fibres, the effect of hot pressing at 3000 lb in^{-2} (21 MN m^{-2}) at 650°C for 10–20 min was to reduce this to 24. A large increase in the spread of fibre orientation also occurred during pressing. These two effects are to a large extent related and both contribute to considerably reducing the axial strength of the composite. To a large extent the problem would appear to be caused by the large particle size aluminium powder used, compared with the whisker diameter since this would prevent intimate mixing of whiskers and matrix and result in fibre to fibre contact. In addition the presence of the oxide skin on the aluminium particles means that even close to the melting temperature of aluminium, higher pressing pressures are needed.

By using liquid phase hot pressing with a silicon carbide whisker–aluminium silicon alloy system, more encouraging properties were obtained.[172] In this case pressing is carried out at a temperature between the liquidus and solidus, and by using careful temperature control it was found possible to liquid phase hot press with alloys having as little as 2·5% silicon. It was found that strengths of nearly 80×10^3 lb in^{-2} ($\simeq 552$ MN m^{-2}) were obtained at a fibre volume fraction of 30% although no indication of fibre damage or comparison with predicted properties is given.

CONCLUDING REMARKS

Studies of the forming processes for metal matrix composites are still relatively limited, principally because they have not yet achieved technological or commercial significance. Clearly an upsurge of work in this area will result once potentially attractive systems are identified.

However, it is hoped that this brief review indicates the nature of the various problems that arise during the fabrication of composites of this kind.

9

Sintering of Metals and Alloys

INTRODUCTION

A distinction has been made in metal sintering practice between American and European techniques. In the U.S.A. the great majority of iron-based sintered parts are produced by a single sintering operation of an hour or less at temperatures not exceeding 1150°C. The European technique involves double pressing and double sintering using higher pressing pressures and higher sintering temperatures. However, the two techniques are no longer confined to their original localities.

There are two main items of plant involved in sintering: the furnace with its auxiliary power and control equipment, and the plant for generating the controlled atmosphere. These items are discussed in a later chapter.

This chapter discusses the effects of the major process and material variables on sintering. In addition, an important aspect of sintering practice is dimensional control and the factors affecting this are discussed. The main advantage of powder metallurgy lies in the production of precision components requiring little or no machining.

PROCESS VARIABLES

The most important factors involved during the sintering process are temperature, time and furnace atmosphere. The influence of these factors on the operation and economy of a sintering plant are discussed below.

Sintering Temperature

The temperatures used in commercial sintering practice may be below 950°C (when sintering brass, bronze or similar materials), 950–1150°C (for iron powder metallurgy) or above 1150°C (for special applications). The upper limit for medium temperature sintering is set by the fact that the continuous mesh-belt conveyor furnace cannot be used at higher temperatures than 1150°C. The possibility of using mesh-belt conveyor furnaces for sintering is important from many points of view. This type of furnace offers in general the best economy and large quantities of parts can be sintered in more uniform conditions than if they were packed in sintering trays.

In certain cases the use of sintering temperatures above 1150°C can be avoided by the addition of suitable alloying elements to give higher mechanical and physical properties. However, where high tensile strength is combined with a demand for high ductility, or increased diffusion is required, then sintering temperatures above 1150°C are essential.

The distribution of temperature in a continuous sintering furnace is important in controlling the quality of the products. Pressed parts contain lubricant which is added to reduce die-wall friction. If removal of this lubricant is carried out in the sintering furnace, then the temperature distribution of the de-waxing zone must be such that no sintering takes place before all the lubricant has evaporated.

In order to run the sintering furnace at its maximum production capacity, the sintering charge should be at the desired sintering temperature for the longest possible portion of the length of the heating zone. The sintering temperature should be reached in the pre-heat zone and this temperature should be maintained until the compacts enter the cooling zone. A pre-cooling zone is therefore advisable to prevent thermal shock between the heated zone and the cooling zone. Rapid cooling should be avoided in iron parts subsequently sized or coined because hardening may occur. Furthermore, if the temperature of the cooling zone is below the dew point of the atmosphere then condensation can occur producing parts that are discoloured.

Sintering Time

Although the degree of sintering increases with increasing time the effect is small in comparison to the temperature dependence. The output of a sintering furnace is inversely proportional to the sintering time. Therefore, an attempt should be made to achieve the desired properties of the sintered parts by shorter sintering times and correspondingly higher temperatures. However, the maintenance costs and energy consumption of a furnace increase when its operating temperature is raised.

Atmosphere

The proper production, use and control of sintering atmospheres which is essential for the optimum use of the powder metallurgy process is discussed later.

MATERIAL VARIABLES

Particle Size

In terms of the basic stages of sintering, decreasing particle size leads to increased sintering. The smaller particle size has a greater pore–solid interfacial area producing a greater driving force for sintering. It also results in a larger inter-particle contact area for volume diffusion, probably a smaller grain size promoting grain boundary diffusion and a greater surface area which might mean more paths for surface diffusion transport.

Particle Shape

Factors that lead to greater intimate contact between particles and increased internal surface area promote sintering. These factors include decreasing sphericity and increasing micro- or macro-surface roughness.

Particle Structure

A fine grain structure within the original particles can promote sintering because of its favourable effect on several material transport mechanisms.

Particle Composition

Alloying additions or impurities within a metal can affect the sintering rates. The effect can either be deleterious or beneficial depending upon the distribution and reaction of the impurity.

Green Density

A decreasing green density signifies an increasing amount of internal surface area and, consequently, a greater driving force for sintering. Although the rate, at any time, and the percentage change in density, increase with decreasing green density, the absolute value of the sintered density remains highest for the higher green density material.

DIMENSIONAL CHANGES

Changes in dimensions resulting from sintering represent an extremely important area in powder metallurgy, especially with respect to large-scale production of parts with small dimensional tolerances. The fundamental process of sintering leads to a reduction in volume because of pore shrinkage and elimination. However, this shrinkage effect may be significantly affected by other causes.

Entrapped Gases

The expansion of gas in closed porosity has been postulated as producing compact expansion. However, the most likely mechanism involving plastic deformation of the surrounding solid is probably insignificant in most cases.

Chemical Reactions

The reaction of hydrogen in the sintering atmosphere and oxygen in the compact producing water vapour can lead to compact expansion. The same result can occur due to oxidation in a poorly controlled atmosphere. Compact shrinkage may occur as a result of reactions leading to the loss of some element to the atmosphere.

Alloying

Alloying that may take place between two or more elemental powders very often leads to compact expansion. This effect which is due to the formation of a solid solution is often offset by shrinkage of the original porosity. Dimensional changes may also occur in a binary system where the rate of diffusion of each metal into the other is different.

Shape Changes

Compacts invariably contain variations in green density. Such variations can lead to substantial changes in shape because of the strong dependence of sintering, especially shrinkage, on green density.

LIQUID-PHASE SINTERING

The three well-established criteria for such a system are:

1. an appreciable volume of liquid phase;
2. an appreciable solubility of the solid phase in the liquid;
3. complete wetting of the solid phase by the liquid phase.

The sintering of hard metal alloys is a well-established application of liquid-phase sintering. Because of the great hardness of the carbide particles which constitute the bulk of a hard metal powder it is quite impossible to press the powders to a density higher than about 60% of theoretical, yet on sintering a perfect pore-free compact can be obtained. Considerable contraction occurs on sintering, typically some 20% on all linear dimensions.

The detailed metallurgical description of the sintering operation is made most complex by the complicated nature of the phase equilibria involved. However, very simply it can be said that sintering starts by solid phase diffusion, frequently aided by a carburizing atmosphere, resulting in the formation of small quantities of W–Co–C eutectic at suitable points of contact in the compact. This eutectic has a melting point of ~1285°C (1560 K). Once small amounts of liquid have formed, its further formation is rapid as surface tension forces cause rapid flow and wetting of all the other solid surfaces. This is accompanied by considerable contraction, often 40% by volume. Perhaps the most remarkable feature of the whole operation is the great speed at which densification occurs at only a few degrees above the eutectic temperature. Quite large compacts can be seen to contract completely in a few seconds. Complete densification will occur with very small quantities of liquid phase, much less than might seem reasonably described as appreciable, about 1% by volume being sufficient to give very rapid and near-complete densification. This must obviously take place by considerable rearrangement of the position of the carbide particles to give maximum packing density. However, such small volumes of liquid are concerned that complete densification could not be possible by particle rearrangement alone. Even with the best packing of the solid particles the liquid phase would be insufficient to fill completely the interstices and the compact would be far from completely densified. Some changes in particle shape must also occur and there must be considerable material transfer at contact points in order to bring about this shape change. In this system the only possible process that could bring about such a shape change would be the solution–reprecipitation process which in this system must occur with great rapidity.

With hard metals sintering treatments are used not only to obtain a dense compact but also to give some control over mechanical properties. Thus in practice sintering times and temperatures are substantially greater than those required to give complete densification. These treatments are given to allow some controlled grain growth during sintering. Considerable work has been carried out on this subject. From a purely physical metallurgy viewpoint the factors involved are composition, average grain size and grain size distribution. In practice grain growth is very sensitive to slight changes in carbon content, and hence all the factors described in the section on sinter furnace atmosphere play a big part. In general, hydrogen atmosphere furnaces are carburizing whilst vacuum furnaces are not. Thus for similar thermal cycles hydrogen sintering generally gives a coarser-grained softer product. This is responsible for a commonly held view that 'vacuum sintering gives a harder product than hydrogen sintering'. Naturally, by suitable adjustments the same products can be obtained from both types of furnace.

In the production of sintered components the presence of a liquid phase during all or part of the sintering cycle represents a unique situation. Liquid phase sintering usually produces a very high density sintered material with little residual porosity. It can produce a desirable metallurgical structure giving excellent mechanical properties. If the identity of the liquid phase is retained in the sintered material, then it often contributes some unique property to the composite.

A process included in the category of liquid phase sintering is the case of liquid infiltration combined with sintering. In this process a mass of a lower melting point metal is allowed to melt and flow into the porosity of a green compact. Both types of liquid phase sintering lead to the attainment of high sintered densities. However, the infiltration technique achieves densification without necessarily causing any shrinkage of the original green compact.

MICROGRAIN CEMENTED CARBIDE

Micrograin cemented carbide has been widely publicized in America since about 1968. The term seems to mean different things to different manufacturers, but in general it is found to be WC–Co alloys in which the average grain size is below 1·0 μm, and sometimes as fine as 0·7 μm. Some advertising claims also suggest that additions are made to dispersion-harden the binder phase.

However, the additions are generally vanadium, chromium or tantalum carbide. It is well known that these have a powerful grain growth inhibiting effect when sintering hard metal. Such fine grain products are very difficult to sinter without rapid grain growth and this inhibiting effect would seem to be the true role of these additives.

Patents[173] indicate involved chemical methods for producing ultra-fine tungsten carbide by reducing WO_3 dissolved in molten calcium chloride with calcium and carbon at 750°C. It is claimed that a pure WC grain with size of 0·02–0·05 μm is produced which is then intensively milled with cobalt. This ultra-fine grain product is extremely sensitive to grain growth on sintering, so consolidation is by a rapid high-pressure low-temperature hot-pressing procedure (one minute at 1400°C at 4000 lbf in^{-2}). This procedure is also claimed to give a preferred orientation effect in the WC grains. In practice the material as hot-pressed has a grain size of 0·7–0·8 μm and shows no detectable preferred orientation and all the common defects of hot-pressed material, especially cobalt segregation.

Other American products described as 'micrograin'[174] are clearly made by the normal carbide production methods using rather fine tungsten metal as the starting point. Such materials have been well known in Europe since the mid-1930s when such a product was introduced in Germany. These alloys have a high combination of wear-resistance and strength, when strength is judged as the conventional transverse rupture strength. However, it is well known that this measure gives little or no guide to toughness, and these materials have a very poor toughness. In Europe they quickly settled down into use solely as a fairly low-cobalt (5–7%) grade intended to give the ultimate in abrasion resistance. The higher-cobalt grades now being offered in America were found to have good resistance to abrasion and attrition wear but to be very unreliable because of their excessive brittleness. However, they do have a few special applications where attrition wear is a serious problem. As cutting tools they may well function[175] better than high-speed steel at the low cutting speeds frequently associated with multispindle automatics. The most likely application in other spheres may prove to be in tools for blanking, pressing and perhaps cold extrusion.[176]

METAL-BONDED DIAMOND TOOLS

Few authoritative papers have appeared on the subject for, typical of a relatively young industry, intense secrecy is adopted as a safeguard.

Theoretical considerations have been made by Jones[177] who considers the requirements for expansion, thermal conductivity and damping. In practice such considerations are normally secondary to abrasion resistance and diamond retention, although in turning tools abrasion resistance is of minor importance and the stone is frequently pre-shaped to aid retention. In this case a high damping capacity is desirable.

Surface set tools, e.g. mining bits or rotary dressers, utilize large stones compared with the 16/18 B.S.S. and finer powders used for impregnated tools in which diamond is distributed randomly in depth. The former require a matrix in which both abrasion resistance and diamond retention are at a maximum. To this end, a tungsten-based matrix is universally employed. Manufacture is achieved by locating the diamond on the walls of a graphite mould with adhesive, packing and consolidating tungsten powder and subsequently infiltrating with a copper-based alloy at temperatures between 900 and 1200°C. Diamond-holding properties produced by this means are reputed to be enhanced by reaction of the diamond with tungsten which is supported by the fact that the graphitized layer invariably shows a tungsten content. The value of such a reaction in terms of the mechanical strength of the interface is, however, dubious.

In the case of impregnated tools the matrix wear rate must be balanced to equal that of the diamond. A range of alloys is used to achieve this, varying from cobalt-cemented carbides to Epsilon bronze. In recent years an increasing proportion of this type of tool is being made by infiltration techniques using tungsten, tungsten carbide and cobalt in varying combinations with lower melting copper-based infiltrants, frequently cupro-nickel and either free-standing or in a graphite mould. Techniques of hot pressing or pressure sintering as outlined by Chalkley and Thomas[178] still, however, predominate.

10

Technology of Pressure-assisted Sintering of Ceramics

INTRODUCTION

Pressure-assisted sintering (PAS), often less precisely called hot-pressing or pressure-sintering, is the simultaneous application of pressure and heat to a powder mass enclosed in a die. In general, the technique allows the use of lower temperatures and pressures and shorter processing times than those for cold-pressing and sintering, and can assist in the production of bodies having a finer grain size, lower porosity and higher purity. In addition the die gives precise dimensional control. Hoyt's[179] description of a PAS method for the production of cemented tungsten carbide is among the earliest references to the process. The technology received impetus in Germany during the 1939–45 War[180] again in the preparation of carbides. Since then it has proved invaluable in the production of a wide range of ceramics and intermetallics. One of the first systematic studies was carried out by Murray *et al.*[40] and Warman and Budworth[81] on several ceramic oxides.

THE BASIC PROCESS

In its simplest form the technique requires a refractory die, a pressure source, a heat source and a temperature-indicating device. The die consists of three components: a thick-walled cylinder, the die-body; a plug or bottom punch, which is inserted in the lower end of the cylinder; and a top punch. For many applications the whole die assembly and its contents are heated, either by a separate furnace or by inductive or resistance heating of the die-body itself. Uniaxial pressure is applied through the ram of a hydraulic or pneumatic press, or by a loaded lever. Temperatures are usually measured by a thermocouple or pyrometer. Simple, laboratory-type PAS units have been described in the literature.[40,182,183]

THE MATERIALS AND DESIGN

The criteria for selection of a suitable material for a die are as follows:

(a) adequate strength and creep resistance under load at temperature;
(b) chemically unreactive;

(c) low thermal expansion coefficient, i.e. lower than the material being compacted, otherwise hot-ejection will be necessary to avoid cracking the sample—or splitting the die;
(d) oxidation resistant;
(e) good thermal-shock resistance;
(f) hard, wear-resistant;
(g) available over a practical size range;
(h) machinability.

Table 6 shows a range of materials that have been used or considered for PAS dies. Graphite is the most common material used in the construction of die assemblies and Campbell and Ford[184] have discussed its capabilities and the criteria for die design. Compacting pressures are limited to about 5000 lb in^{-2} (~34 MN m^{-2}) up to 5 cm i.d. and to about 2000 lb in^{-2} (~14 MN m^{-2}) at 15 cm i.d. Spriggs *et al.*[185] and Haertling[186] report using alumina as a die material; the former to obtain higher pressures in the region of 1400°C, the latter to avoid contamination and reduction in the fabrication of ferroelectrics. Aluminium nitride has been employed in the manufacture of hard metals and reducible oxides.[187] Moss and Stollar[188] discuss the stresses imposed on a die-body and describe the design of a graphite liner bolstered in compression, both radially and axially, with a Mo alloy. Square-section dies (Sambell, unpublished work) can be built up from flat graphite plates supported by refractory bolsters, strapped together with a water-cooled yoke or spring-loaded band.[189] The poor wear property of graphite has led to several die designs incorporating replaceable sleeves. Sambell[190] has avoided ejection problems by using a multicomponent assembly within the die body.

TABLE 6. Uniaxial PAS die materials

Die material	*Maximum use temperature* °C	*Maximum pressure* kpsi (MN m^{-2})	*Comments*
Graphite (commercial) Acheson CS Grade Morganite EY Series	2500	5 (~34·5)	Protective atmosphere required above ~1300°C; exceptionally, used up to 3000°C.
Graphite (specials) Poco	2500	20 (138)	
AlN	1200	15 (104)	Expensive; obtainable only in small sections; difficult to machine; very brittle; oxides least reactive but creep limited.
Al_2O_3	1400	30 (207)	
BeO	1000	15 (104)	
SiC	1500	40 (276)	
TaC	1700	8 (~55)	
WC, TiC	1400	10 (69)	
TiB_2	1200	15 (104)	
W	1500	3·5 (~24)	Easily oxidized; tendency to gall
Mo, Mo alloys	1100	3 (~21)	
Nimonic alloys Immac. 5 stainless steel	110	Varies	Galls easily; creep limited.

PRESSES

Hydraulic presses are the most frequently used source of pressure because of the degree of control that can be obtained. Necessary design features are large platen areas and adequate daylight between the platens. Naiguz and Mil'shtein[191] give a detailed description of a 40 ton capacity press with the heat source integral with the press platens and capable of reaching temperatures of up to 2800°C. Other methods of load application have been used, the most common of these being the lever/weight system.[183] Stringer and Williams[129] emphasize the preference for this method in their reaction PAS technique and describe single and double action facilities.

FURNACES

PAS furnaces are usually of vertical construction, although horizontal furnaces are described by Johnson[192] and Zehms and McClelland.[193]

Resistance wire or tape elements can be used up to 1800°C. A resistance-heated, graphite tube has also been used as a furnace.[40]

The use of graphite as a die material allows the die itself to act as a furnace, either by induction or resistance heating. Wolff[194] has described an apparatus in which the die is located inside an evacuated silica tube. An induction coil is used externally and temperatures up to 2500°C have been reached at a vacuum pressure of approximately 1×10^{-3} Torr. Stringer and Williams[129] include a metal vacuum chamber in their description of a simple PAS unit.

TEMPERATURE MEASUREMENT

The most common devices are thermocouples and optical or radiation pyrometers. The importance of measuring the correct temperature and obtaining reproducibility cannot be stressed too highly. Thermocouple hot junctions are susceptible to degradation and techniques for their protection are described by Desport,[195] Auskern and Thompson.[196]

MECHANICAL AIDS TO PAS

Single-ended, uniaxial compaction of refractory ceramic powder systems leads to excessive pressure variations in the compacted powder, particularly along the length. This causes serious variation of sintered density above a length-to-diameter ratio of about 1·5, thus seriously limiting the application of PAS. The problem is only slightly eased by using double-ended compaction or a 'floating' die. This restriction has led a number of investigators to develop other methods of densification, as follows.

Isostatic Pas

Hodge[197] exploited this technique in the ceramics field and discusses design principles and more recently Long and Snowdon[198] have described improved designs.

The technique involves evacuating and sealing a fine powder in a metal can, which is then autoclaved. Hodge reported achieving up to 99·9% of theoretical density in oxide ceramics at temperatures between 1150 and 1400°C with pressures of 10–20 kpsi (~70–140 MN m^{-2}) for 30–180 min. Practical applications have been reported by Ryan and MacMillan,[199] Huffadine *et al.*[200] and Egerton and Bieling.[201] Some patent literature on techniques is given in Refs 202–204.

Zone Pas

Oudemans[205] has perfected a continuous uniaxial PAS process, which is used in the production of long rods of ferroelectric ceramics and gives material of uniform density and grain size, with a high surface finish.

Applied Metallurgical Processes

Processes that have been adopted to produce large-aspect ratio artefacts are hot-extrusion, hot-rolling and hot-forging. Rice[206] gives a full account of these techniques in his review of hot-pressing processes. In each case high-quality ceramics have been obtained. Spriggs *et al.*[207] have described a 'press-forging' process that is used to produce high-strength ceramics from PAS billets, which are dense but weakly bonded. Working pressures up to 100 lb in^{-2} (~700 MN m^{-2}) and temperatures up to 1500°C have been used to produce fully dense oxides.

High-Energy Rate Fabrication

Attempts to reduce the PAS time-cycle have led to investigations of high-energy-rate compaction processes. Leonard[208] and Porembka[209] have explored the possibilities of detonating explosive charges adjacent to a powder mass. Hallse[210] has employed a high-velocity pneumatic–mechanical ram (Dynapak) to produce high-density Al_2O_3, LiF and clays. In both techniques the powder charge is pre-heated and should preferably be sealed in an evacuated container. Very high compaction pressures of brief duration (several microseconds) are achieved and offer high production rates.

High Pressure and Temperatures

The application of ultra-high pressures at high temperatures, stems directly from the technology associated with the synthesis of diamond, which is perhaps the most dramatic demonstration of the potential of this technology. Pioneering work was carried out by Hall[211] and later by Daniels and Jones[212] with the development of the 'belt' in which materials can be subjected to 1·5 million lb in^{-2} (~10 GN m^{-2}) at temperatures in excess of 2000°C over a period of hours, if necessary. The technique has been applied to ceramics by Hall,[211] Delai *et al.*[213] and Kalish and Clougherty.[214] The technique and die designs have been reviewed by Schwartz.[215]

Vacuum Pas

The elimination of physisorbed and/or chemically-bonded gaseous impurities during the compaction process has been shown to be a very important factor in the subsequent high-temperature performance of ceramics. Rice[216] has investigated the causes leading to pressurised, entrapped gases. These gases can produce blistering, delamination or explosive disintegration during the subsequent heat treatment of high-density materials. Miles *et al.*[217] showed the advantages of using vacuum PAS to remove residual gases, so achieving transparency as distinct from translucency in the fabrication of magnesia.

Continuous or Repetitive Pas

Finally, some continuous, automatic and semi-automatic processes that have recently been published in the open and patent literature. Reference has already been made to Oudemans' process,[205] and the numerous metallurgical processes that have been adapted by Rice[206] are basically continuous processes. Alliegro and Foster,[218] have recently described an automatic process which is resistance-heated to 2500°C. The facility has run in uninterrupted spells for 1000 h, producing 24 boron carbide tiles ($150 \times 150 \times 10$ mm) per hour.

Meadows[219] describes a repetitive PAS process in which the powder is contained in a series of re-useable shells, each complete with pistons. The whole process, from powder feed to ejection, is cyclic and the apparatus operates *in vacuo* or in an inert gas up to 2500°C and 30 klb in^{-2} ($\sim$200 MN m^{-2}). The semi-automatic processing of BeO by Zehms and McClelland[193] is based on a chain of dies that are continuously fed into a horizontal furnace, the compacting pressure being transmitted from one die to the next.

THERMOCHEMICAL AIDS TO PAS

Despite the many mechanical aids that have been developed, the temperature and pressures required to produce high-quality ceramics are still uncomfortably high. Several techniques are evolving that make use of chemical reactions and low viscosity phases in order to reduce the conditions required to obtain particular properties.

Viscous Phase Pas

The addition of glassy phases to crystalline ceramics provides a relatively low-temperature viscous phase that assists particle movement and eliminates porosity. The glass remains *in situ* at the grain boundaries and tends to dominate the high-temperature properties of the material.

Transient Liquid Phase PAS

The advantages to be gained from a chemically induced, transient liquid phase are exemplified in the production of fully dense, transparent MgO, at moderate temperatures and pressures. Rice[216,220] pioneered this work, which involves adding 1% of LiF powder to the MgO. During the process a volatile eutectic is formed in the grain boundaries which liquefies at ~760°C.[217] The LiF is almost completely eliminated by post-firing at 1200°C.

In a process developed by Stoops[221] a liquid phase metal is used to enhance densification. The metal is progressively extruded from the artefact; provision being made in the die design to collect the metal.

Reactive PAS

Chaklader[222] has recently reviewed the work carried out in this field. The word 'reactive' is used to cover various decomposition, dissociation and phase inversion processes. Examples of decomposition and dissociation are provided by the work of Morgan and co-workers,[223,224] where fully dense oxides have been obtained in the temperature range 900–1200°C at typically 10 lb in^{-2} (~70 MN m^{-2}). In addition, Chaklader and co-workers[123,222,225,226] have used phase transitions to induce densification at relatively low temperatures and pressures. However, in the case of α-Al_2O_3, Matkin *et al.*[98,227] observe a limiting density, beyond which it is necessary to go to higher temperatures. Carruthers and co-workers[124,228] have also used the technique to obtain high-strength clays.

Reaction PAS

The term covers techniques making use of chemical reactions between individual components of a ceramic. Stringer and Williams[129] have utilized the highly exothermic reactions that occur when elemental powders are heated together. Pressure was applied to metal–boron powder mixes at a critical, transiently plastic stage. Stephens and Hoyt[229] have used a mixture of ZrC, UO_2 and graphite in order to produce solid solution ZrC–UC. Accary and Caillat[230] have reacted uranium with nitrogen gas or carbon during hot-pressing to produce UN and UC. Okamoto *et al.*[231] have combined reactive and reaction hot-pressing techniques in the fabrication of nickel–ferrite bodies.

SUMMARY

The wide range of mechanical and thermochemical techniques that are being developed serves to emphasize both the desirable qualities of PAS ceramics and the individuality of many of them. The marked improvements in quality and reliability are reflected by the number of manufacturers who are now marketing hot-pressed components. Table 7 gives an indication of the materials that are worked with and the range of applications.

TABLE 7. Companies using PAS for the fabrication of ceramics

U.K. and Europe	
Consolidated Borax	TiB_2, ZrB_2, B_4C, SiC, BN, Si_3N_4 Boride–graphite and boride–BN cermets
Carborundum	see U.S.A.
Lucas	Si_3N_4
Allmanne Svenska	Oxides
Electriska Aktiebologet (Sweden)	Carbides
Deutsche Edelstahlwerke Aktiengesselschaft (Germany)	Oxides
Philips (Holland)	Ferroelectrics
U.S.A.	
Avco	HfB_2, ZrB_2, B_4C, Al_2O_3
Carborundum	TiB_2, ZrB_2, $ZrB_2/MoSi_3$, B_4C, HfC, TiC, ZrC, AlN, Al_2O_3
Kodak	CaF_2, MgF_2, MgO, ZrS
National Beryllia	ZrB_2, BeO, BeO/SiC, Al_2O_3, ThO_2, ZrO_2, Zr silicate
Norton	ZrB_2/SiC, B_4C
Union Carbide	BN
V. R. Wesson	Al_2O_3

Applications: reaction vessels, liners, pouring spouts, spray and shot-blast nozzles, air-bearing components, nuclear control rods, electrodes, crucibles, tool tips, dies, infrared optics, turbine blades, thread guides, wear plates, valve seats, re-entry nose cones, armour, ferroelectrics.

11

Technology of Pressure-assisted Sintering of Metals

INTRODUCTION

The scope of the problem as applied to ceramics has been discussed in the previous chapter and the same general considerations apply to the pressure-assisted sintering of metals.

The properties that can be obtained by conventional powder metallurgy techniques are clearly limited due to the presence of porosity. In the production of sintered components the amount of densification is controlled in order to produce very close tolerances. Therefore, the porosity (about 15%) which is restricted by economic tableting pressure remains in the final sintered compacts. The important effect of density on the mechanical properties is illustrated in Figs 8–11 for iron powder (given by Squire[232]) and emphasises the need for high densities if mechanical properties approaching those of bulk metals are needed. In addition, there are problems associated with powders which are difficult to compact at room temperature and also applications where the presence of porosity cannot be tolerated.

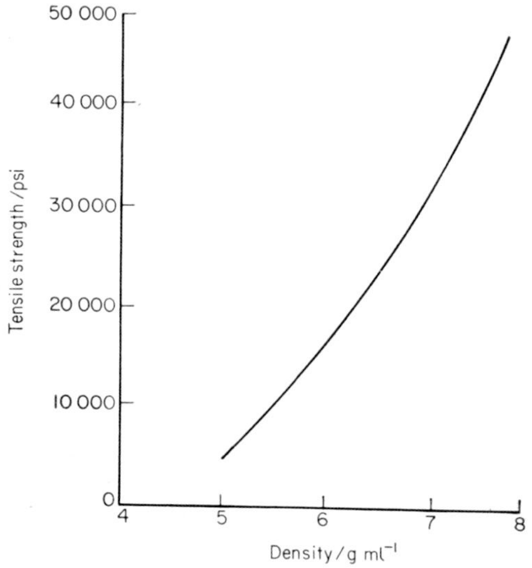

Fig. 8. The effect of density on tensile strength for sintered iron.

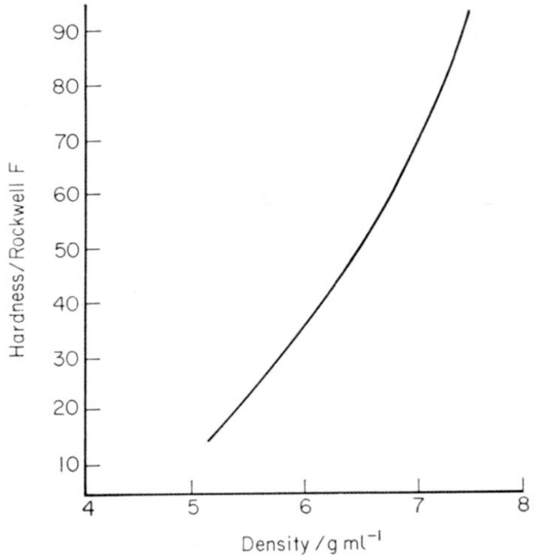

Fig. 9. The effect of density on hardness for sintered iron.

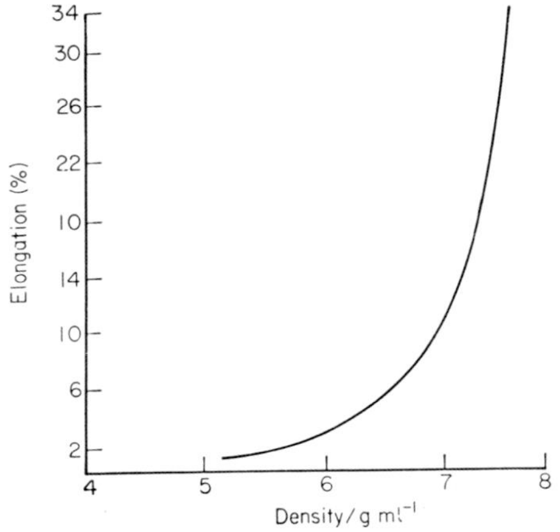

Fig. 10. The effect of density on elongation for sintered iron.

In order to overcome these problems various techniques involving pressure-assisted sintering have been developed. These include hot pressing of powder in a die, hot isostatic compaction, powder forging and powder extrusion.

HOT PRESSING

In cold pressing, densification is increasingly retarded due to the effects of work hardening within the powder particles. So much so that it is impossible fully to densify most metal powders by pressing at room temperature. However, if the powder compact is maintained above the recrystallization temperature the effects of cold work are eliminated. In addition it is possible by hot pressing to form dense compacts from materials which at room temperature display little or no plasticity.

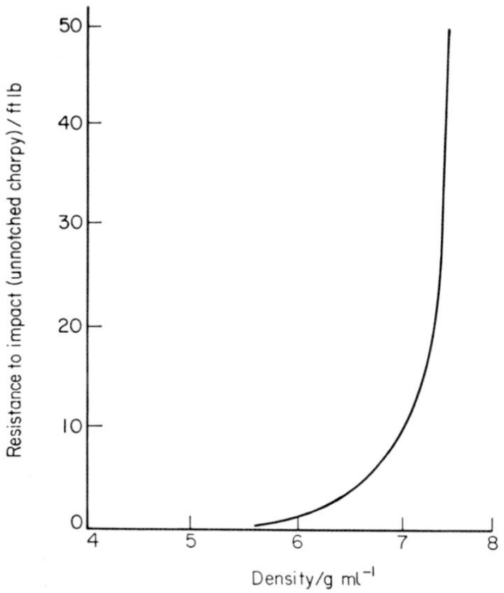

Fig. 11. The effect of density on impact resistance for sintered iron.

Early work on hot pressing has been reviewed by Goetzel[1] while Williams[233] has studied the effects of load, time and temperature on the densities of various metal powders.

While the hot pressing technique has important advantages, outlined above, there are definite practical difficulties and disadvantages. For most metal powders the first problem is that of preventing oxidation of the powder during heating, pressing and also during cooling after ejection. The choice of die materials is limited at elevated temperatures. High-speed steel dies will operate satisfactorily only up to 500°C. Above 500°C it is necessary to use nickel or cobalt superalloys or cemented carbides. Above 1000°C the field is limited to graphite, oxides and possibly nitrides, borides or silicides. Of these graphite is the most popular because it is cheap, easy to machine and has excellent thermal shock resistance. It may be used up to 3000°C and being an electrical conductor, it allows direct heating of the die by either resistance- or induction-heating. The disadvantages associated with the use of graphite are possible reaction with the material being pressed and a load limitation of about 2 tonf in^{-2}.

For most metal powders the hot pressing temperatures are below 1000°C and metallic dies are usually adequate. However, from a practical viewpoint hot pressing does present a greater problem with respect to dimensional control and the large scale production of parts, especially in comparison with the more conventional approaches.

Pressure-Assisted Sintering as Applied to Hard Metals

The cemented carbide industry has probably made greater commercial use of hot pressing than all other industries put together. The reason for this is primarily economic, technical factors having played only a small part. The hot press method has been uniquely satisfactory for the manufacture of large pieces of cemented carbide for such purposes as very large dies, compacting moulds, deep drawing rings, rolls, etc. The necessary mould is easily and cheaply made in graphite and a simple hydraulic press and medium frequency generator of about 50 kW capacity is ample for a wide range of sizes. Pieces of cemented carbide up to 24 in diameter and over 100 kg in weight have been commonly made in this way.[234] The alternative manufacturing method using large steel dies for cold pressing and very large batch type furnaces for pre-sintering and final-sintering is hopelessly uneconomic in comparison. However, the increasing use of isostatic pressing and large vacuum furnaces in recent years has greatly influenced the position and there has been a substantial decline in the amount of hot pressing.

HOT ISOSTATIC COMPACTION

The process of hot isostatic compaction has been mainly developed to meet the need for producing non-porous components of materials, especially composites, which are difficult to sinter. In many cases these components are long and slender, for instance nuclear fuel elements, whose forms rule out hot-die compaction. In other cases the stress and temperature requirements are such that hot pressing in graphite dies is not possible.

To meet these needs the type of equipment used for gas pressure bonding has been found suitable. However, earlier attempts at hot isostatic compacting were made using simpler equipment such as the hot wall apparatus. This type of apparatus is clearly limited to fairly low-temperature use by the creep strength of the pressure vessel material. Other earlier attempts utilized liquid metal as the pressure-transmitting medium, but this was limited to modest operating temperatures.

For temperatures above 1100°C, the range of most interest, the cold wall equipment developed at Battelle has been most used. These pieces of equipment use either argon or helium to transmit the pressure and much care is required to contain the gas or convection effects which can lead to very large temperature gradients and rapid furnace failure.

Preparation of specimens is usually a fairly lengthy and expensive procedure. The powders are generally first compacted to the highest possible density and machined to give a form which is easily encapsulated to produce a pre-form which will give the required final shape. Sintering is often carried out either prior to or after machining to thoroughly de-gas the compact.

Encapsulation is the final step in specimen preparation. Thin-walled metal cans are used, the metal needing to be compatible with the compact at the hot-pressing temperature. It is usual to electron-beam weld the can in order to ensure that a minimum quantity of gases remain inside the capsule. For temperatures of about 1200°C stainless or mild steel cans are used with tantalum finding use at higher temperatures.

After hot isostatic compaction the can must be removed from the compacts either by machining for simpler cylinders or by pickling.

Hot isostatic compaction has been used for a wide variety of materials. Mainly these have been ferrous and refractory metal base composites containing fissile oxides or carbide dispersions, refractory metal alloys, and high melting point oxides and carbides.

At present much of the equipment described above is insufficiently reliable for other than research and development purposes and is undoubtedly expensive to operate.

Isostatic Hot Pressing of Cemented Carbide

During the last five years isostatic hot pressing has become so firmly established in the cemented carbide industry that it is probably the world's largest user of isostatic hot-press equipment. At first this might seem surprising as it is well known that cemented carbides readily sinter to near full density whilst isostatic hot pressing has become widely recognized as the ultimate treatment for 'difficult to sinter' materials.

However despite the ease with which cemented carbides sinter to near-theoretical density they nevertheless almost invariably show some residual porosity, generally of the order of 0·001–0·01 % by volume, distributed as fairly randomly sized voids mostly in the size range up to 25 μm, with a few much larger voids. This porosity is a serious problem because of its effect both on surface integrity and upon strength.

Many applications for hard metal require large polished areas of high surface integrity completely free from even the smallest pore. Typical tools are bar- and wire-drawing dies, cold rolls, cold heading tools, extrusion punches and metal powder-compacting dies. On such tools even tiny pores below 25 μm in diameter can ruin the surface finish of the component being produced and give rise to rapid wear that causes accelerated deterioration in surface finish and poor tool life. The presence of these flaws cannot be detected by any satisfactory non-destructive method before tools are polished and ground so they are not found until the tool is completely finished, after expensive tool making operations. Generally such tools are large and heavy, so the initial cost of the carbide blank will be high, but the cost of grinding and polishing before the fault is discovered could be an order of magnitude higher. Hence there is a very large economic gain to be made by positively eliminating any possibility of such porosity.

It has been known for many years that the porosity of cemented carbide can be considerably reduced by hot pressing as compared with the more normal cold press and sinter method. Although the process of hot pressing in graphite dies has many disadvantages and limitations it was nevertheless frequently used in the manufacture of large components of simple shape that needed to be specially free from porosity. Thus the potential advantages of isostatic hot pressing were apparent but for some time they were unattractive because of the apparently severe encapsulation problems that would be associated with the high temperature required and the reactivity of the material. Eventually it was discovered that cemented carbide, after densification by a normal sintering process, did not require encapsulation for subsequent isostatic hot pressing. At temperatures just above the solidus cemented carbide is a fairly plastic mass with considerable cohesion and surface tension so that it effectively functions as its own envelope during isostatic hot pressing. The only difficulty is that if there are any voids that are connected to the surface or are so close to the surface that the semi-liquid skin is too weak to transmit the gas pressure without rupture, such voids are not closed. The great advantage of this process of isostatic hot pressing after sintering is that it is cheap and needs no special treatment—the pieces have already contracted so that twice as much work can be treated in the same volume as with green compacts and there is no real restriction on size or shape. If a component can be sintered it can be isostatically hot pressed. Once this had been realized it was possible to proceed with practical evaluation tests which have demonstrated the economic advantages of this expensive process.

The other problem resulting from the presence of porosity is the adverse effect upon strength. Cemented carbide is a typically brittle material the strength of which is very largely controlled by the defects that exist in the structure. When a cemented carbide component is loaded, stress concentrations arise at any void, and as the applied load is increased the fracture stress will be reached at such a point long before the average load reaches this value because there is no possibility for the stress concentration to be reduced by local plastic deformation.

Thus the porosity normally present in cemented carbide has a major effect upon its strength, which is largely dependent upon the size and distribution of the pores in relation to the applied stress system. Consequently the strength of any batch of samples of cemented carbide shows considerable variation from one to another and the strength of the batch can only be defined as a statistical distribution. Commonly, in assessing the 'quality' of any batch of cemented carbide undue emphasis has been given to average values whereas the scatter in results is of even greater importance. The tendency for any batch to give premature failures will be associated with the proportion of test pieces that show abnormally low strength. In any cemented carbide application only a low probability of massive failure prior to 'wearing out' can be tolerated.

The elimination of porosity by isostatic hot pressing is found to increase the average strength considerably but a feature of greater practical importance is that the strength of the weakest components is improved to a much greater extent than average strength (Fig. 12).

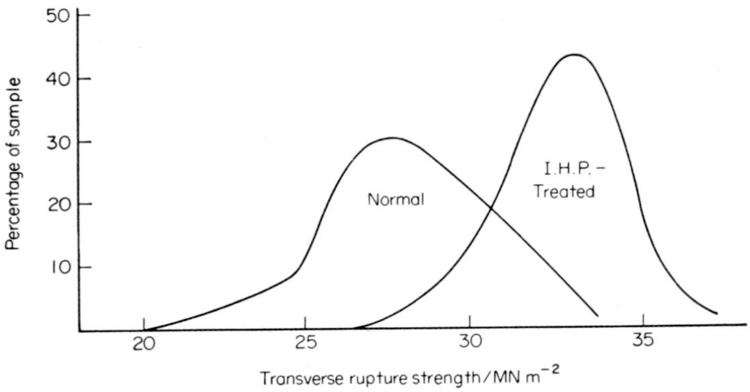

Fig. 12. Transverse rupture strength of rock drilling grade with and without isostatic hot pressing.

The magnitude of the improvement is greatest with the harder and more brittle grades of cemented carbide as clearly these are the grades for which a given void will create the greatest stress concentration. Typical results are given in Table 8 for three different carbide grades.

Although it can be shown that the removal of porosity by isostatic hot pressing will increase the transverse rupture strength it must be understood that this is not necessarily always the case. There are many other defects, in structure, composition and surface condition, that affect the strength of cemented carbide. The effect of all

TABLE 8. Effect of isostatic hot pressing on three different grades of cemented carbide

Grade	*Hardness*	*Transverse rupture strength/MN m^{-2}*	*Coefficient of variation %*
WC 0·8 μm 6% Co	1700 Hv30 Normal	1701	22·5
	i.h.p. treated	3013	11·3
WC 3·5 μm 11% Co	1200 Hv30 Normal	2838	11·3
	i.h.p.-treated	3246	6·3
WC 1·4 μm 25% Co	950 Hv30 Normal	2885	5·6
	i.h.p.-treated	3110	4·4

such defects is not additive, but has a 'weak link'-like characteristic. Thus no improvement will be obtained from isostatic hot pressing if other and more serious defects than the porosity are also present. In many practical applications the most serious effect will be surface defects resulting from sintering, grinding or, most commonly, from wear patterns. With the advent of isostatic hot pressing it will be necessary to pay even more attention to surface finishing techniques if the full advantage is to be obtained.

Structural defects of most types such as η-phase or free carbon can prevent the full strength being obtained, and so can cobalt segregation resulting from undersintering. In an experiment with a 15% Co alloy this effect was shown very convincingly. Table 9 shows results on normally sintered material, seriously undersintered material so that it was very porous, and the undersintered material treated by isostatic hot pressing at the same temperature.

TABLE 9. Effect of isostatic hot pressing on undersintered 15% Co grade

Treatment	*Porosity (ASTM)*	*Average MN m^{-2} Transverse rupture strength*	*Coefficient of variation (%)*
Normal sinter 1400°C	<A1	2600	13·9
Normal sinter 1400°C±i.h.p.	<A1	2950	9·2
Sintered 1300°C	B3	1590	9·4
Sintered 1300°C±i.h.p.	<A1	2200	18·9

In this case the isostatic hot pressing completely removed the porosity but the increase in strength was only small because the strength was dominated by the unsatisfactory highly segregated cobalt distribution due to the low-temperature sintering.

Yet another application of isostatic hot pressing to carbide technology concerns the very hard materials of low cobalt content. For instance it is possible to produce grades for applications such as wire drawing that are free from porosity and of adequate strength with much lower cobalt than is practicable by direct sintering. Already new grades have been introduced for wire drawing. They have a hardness in excess of 2000 Hv and give some 2–4 times the life of the replaced conventional wire die grade.

The greatest obstacle to further increase in the application of isostatic hot pressing of cemented carbides is economic. The equipment required is costly and the financial charges that accrue whether the equipment is used or not are very high. In addition the operating and maintenance costs are not inconsiderable so that the present applications are established in fields where the economic advantages of the high-integrity product are clearly profitable. However, like so many modern processes the cost is very sensitive to output volume, and the cost should drop substantially as the demand increases so that high outputs are obtained from equipment that is kept fully utilized. As user appreciation will develop concurrently with reduction in cost a rapid acceleration in use of isostatic hot-pressed cemented carbide is to be expected.

Other commercial developments have been made recently by two Swedish companies resulting in the A.S.E.A.–Stora Process and a similar development in the United States by The Crucible Steel Co. The former process consists of the production of suitable powder by inert gas atomization. The powder is transferred to a carbon steel capsule and after a lid is welded on it is subjected to cold isostatic compaction at about 4000 atm. The capsule is evacuated, heated to about 1000°C and isostatically pressed at 1000 atm in argon. The resultant product is supposed to be of sufficiently high and improved quality to be worth the high cost. Tool steels have been produced with very fine uniform structures.

POWDER FORGING

The process of powder forging which links the powder metallurgy and forging technologies is increasing in importance and scope and a bibliography has been produced by Hausner.[239] The process involves the production of a suitably shaped pre-form by conventional powder metallurgy techniques using pre-mixed or pre-alloyed powders. The green preform is then either sintered or transferred directly to a suitable furnace in which it is heated to forging temperature under scale-free conditions. The hot preform is transferred to preheated closed impression dies and forged to a density very close to the theoretical wrought value.

A considerable number of process variations exist, depending on the shape of the preform and the tooling system. In certain cases a relatively simple preform may be forged into a complex component producing a large amount of material flow. At the other extreme a complex preform may be forged to produce only densification.

The main advantage of powder forging is that properties equivalent to wrought materials can be achieved, although in this context the powder purity is important. In conventional powder metallurgy products the dominant effect of porosity outweighs any effect of powder impurities. However, for powder forgings, at or near solid density, the effects of impurities become very important particularly with regard to impact toughness and fatigue strength.

Comparing the routes by which traditionally forged and powder forged components are made, the former requires a number of forging blows in a series of dies to develop the final shape. With powder forging only one forging stroke is required in one set of closed dies giving a reduction in actual forging cost and a greater improvement in press utilization. This is effected by the different forging characteristics of a powder preform, the location of metal in the appropriate place, reducing material redistribution, and the absence of flash-reducing the forging load. The improved tolerances gained from powder forging reduce considerably the amount of machining required giving a material utilization of over 90% compared with 40–50% for a conventional forging.

The main components studied so far include gears, drive flanges and connecting rods, the last application being particularly suitable because weight balancing can be eliminated.

The potential growth involved in powder forging is considerable. Sintered parts consume about 9500 tons per annum of iron powder in the U.K., compared with about 700 000 tons per annum of steel used to make small components from forgings. To this figure can be added components made by casting and machining or machining of bar stock to realize the potential scope of powder forging. In contrast, however, the advantages of powder forgings have to be balanced against the increased material cost and this may limit powder forgings initially to particular components.

A recent development in powder–metallurgical forging practice is the Osprey process. This method involves the atomization of scrap by nitrogen and directing the spray of hot particles into shaped moulds. The resultant high-density preforms are subsequently forged.

POWDER EXTRUSION

The hot extrusion process is extremely useful in the production of rods, tubes and similar shapes from powders, since the amount of deformation obtained in one operation is very high and theoretical densities can be obtained. The process has been described by Loewenstein *et al.*,[240] who discuss the extrusion of loose powders, compacted blanks, canned loose powders and powders precompacted and canned or compacted directly into the supported can.

The process may have some difficulty in competing with conventionally produced bars and sections. However, the advantages lie in the structure and properties that can be obtained. The technique has been successfully applied to tool steels producing a uniform distribution of fine carbide particles. This contrasts with the careful and lengthy working cycle necessary to break up the coarse eutectic carbide networks in conventionally produced tool steel.

SINTERED ORE

Sinter was first produced in the early 1930s and some of the initial work has been described by Elliott.[241] Originally the object of sintering was to agglomerate material that was too fine to be charged into the blast furnace directly as it would have caused high dust losses and a burden that was impermeable. The Greenwalt sintering pan has been used,[242] but more commonly the Dwight Lloyd sintering strand is employed. This consists of a series of pallets in the form of a moving continuous belt on to which the feed material, consisting of ore with approximately 40% return sinter fines, 10% water and 3% coke, was charged through a swinging spout to a depth of approximately 15 in. At the charging end is an ignition hood which will raise the surface temperature of the charge to approximately 1300°C. After passing under the hood air is drawn through the bed by a fan which will cause the flame front to pass through the bed at a speed of approximately 1 in min^{-1}. Reaching the discharge end of the strand, the sinter is cooled and screened, the undersize being returned to the sinter strand, the remainder going for use in the furnace. This design is still the basis of modern sintering machines.

The use of sinter increased considerably in the succeeding twenty years and is still increasing as it has been realized that the properties of sinter are superior to those of many ores,[243] and some as mined ore and coke is now crushed to make sinter. In addition to this, many concentrates are now being used in which, as a result of mineral dressing operations, the ore particle size has been reduced to such an extent that it would be quite unusable in a blast furnace directly. United Kingdom practice has been compared with Japanese practice by Dartnell.[244]

At first sintering was only required to make reducible materials of adequately large size that would not break during handling or charging from fine feed stock and the quality of sinter was assessed by a shatter test. It is, however, now known that other factors are important as well as these. As the burden descends through the furnace, the temperature rises and the sintered material is subjected to compressive and abrasive loads at elevated temperatures which can cause a production of fines or collapse of the ore body. If the sinter collapses at elevated temperatures or an excessive amount of fines produced the efficiency of the furnace deteriorates. As the aim is also to reduce the amount of coke used in the furnace, the load on the iron-bearing part of the burden will increase and this has led to investigations of the hot strength of sinter.[245] These tests were originally carried out in neutral or oxidizing atmospheres but it has since been found that in reducing atmospheres similar to furnace atmospheres, the strength changes as the phases present vary, and in addition to this volume changes can occur, and testing therefore has been changed slightly to bring in conditions of abrasion,[246] or alternatively tests are also sometimes carried out to detect the production of fines in reducing atmospheres only.[247] Current British testing practice has been discussed by Davidson.[248]

The emphasis on the use of unwanted fine material for sinter has therefore now been replaced by the deliberate production of fine material, the blending of different types of ores with the aim of introducing into the furnace a feed in which the unwanted constituents H_2O, CO_2 and S have been removed and a slag has formed. It is also required that the sinter should be strong enough both hot and cold, but at the same time it is also required that the reducibility should not be adversely affected. High strength can be achieved by the addition of materials which form a slag bond or alternatively by sintering at elevated temperatures, but unless the bond is itself reducible the reducibility of the sinter will deteriorate. The bonds based on calcium ferrite are currently considered to be most acceptable. The subject has been reviewed by Ball *et al.*[249]

12

Equipment (Ceramics)

CERAMICS

Introduction

The three main kiln types used in the traditional ceramics industries, i.e. intermittent, continuous-chamber and tunnel, were all in use in the 19th century, and the design of the first two had, by the process of much trial and error, reached a high degree of sophistication. But coal was then so plentiful and cheap that it was the only fuel used, and it was this apparent advantage that impeded any further significant progress.

Fig. 13. Open-placed tunnel kiln truck.

Modern methods of firing only became possible in the last half century or so when technology had advanced far enough to provide dependable means of measuring and automatically controlling high temperatures, and also when the development of suitable resistance elements, burners, insulating bricks and fans made it possible to design kilns thermally efficient enough to encourage the use of more expensive fuels such as oil, gas and electricity. No coal has been used to fire pottery since 1967 and the oil, gas and electricity that have taken its place are not only used more efficiently and smokelessly[250] but have also led to a marked reduction in rejects and work-in-progress, considerable labour savings and a great decline in the use of saggars, which were previously used in all direct-fired kilns to support, contain and protect the ware. With the present-day gaseous fuels, which are ashless, smokeless and sulphur-free, the ware requires no protection and is placed on the refractory superstructure of tunnel kiln cars as shown in Fig. 13.

In the brick industry continuous-chamber kilns based on Hoffmann and Licht's original design of 1858 have, from a purely thermal standpoint, been operating extremely efficiently for over a century, and are still coping with the enormous output of the Fletton brick industry.

Figure 14 shows a schematic representation of a continuous-chamber kiln in which four of the sixteen chambers are fired in rotation by introducing fuel through holes in the crown. The hot combustion gases from these chambers are used to heat dry

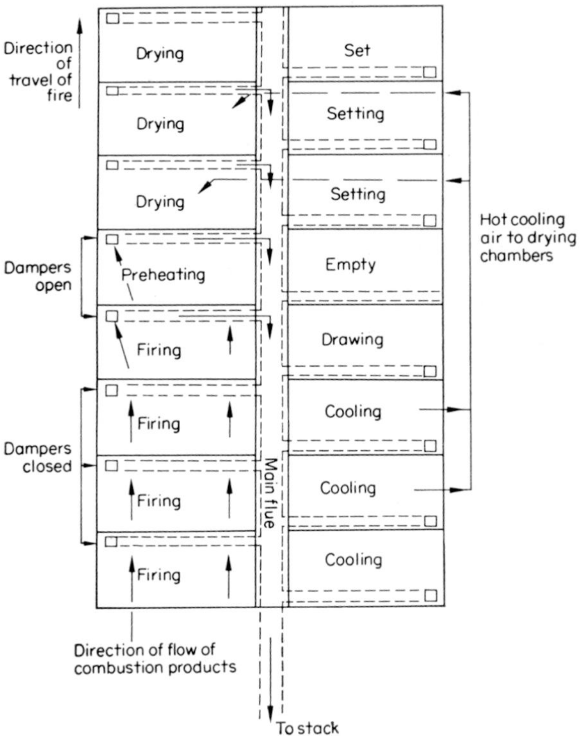

Fig. 14. Diagram of continuous-chamber kiln.

bricks in one or more preheating chambers. Cold air is also drawn through cooling chambers and the resulting hot air is used in drying chambers. Heat recovery, based on the principle of regenerative furnaces, is therefore the chief virtue of this type of kiln.

Tunnel kilns are now widely used in all sections of the ceramics industry with petroleum fuels and gas as the dominant fuels in the refractories and heavy clay industries, and gas as the dominant fuel in the pottery industry with some electricity and petroleum fuels. The tunnel kiln has in fact been the main agent of change and the corner-stone of most modernization plans.

At first sight it appears paradoxical that as the use of intermittent kilns is declining sharply in the heavy clay and refractories industries their use is increasing in the pottery industry. However, in the heavy clay and refractories industries the kilns are traditional, coal fired types whilst in the pottery industry they are of modern design built with refractory insulating brick linings and more recently ceramic fibre linings, and fired by gas or electricity.

The implementation of the Clean Air Act 1956 has accelerated the trend towards modern firing methods and interesting reports dealing with the scheduled processes in ceramics works are to be found in the Annual Reports on Alkali, etc. Works[251] from 1958 onwards.

KILN SELECTION

The choice of kiln employed for firing is influenced by:

(a) type of product;
(b) size of article;
(c) output;
(d) capital and running costs;
(e) flexibility of operation;
(f) production of a high percentage of satisfactory ware;
(g) how it fits in with the present or future production system.

The type and size of product, for example, has a great influence on the choice of kiln since ceramic products are extremely diverse in both size and value. The three main divisions of the industry are:

(i) heavy clay which includes building bricks, roofing tiles, floor quarries, sanitary pipes and conduits;
(ii) refractories which include firebrick, insulating firebrick, ceramic fibre, silica, high alumina and basic refractories;
(iii) pottery which includes earthenware and bone china tableware, wall tiles, electrical porcelain and sanitary ware.

Firing equipment for these products ranges from no kiln at all in the case of stock bricks, which are fired in clamps, to tunnel kilns with multi-zone control. Some articles, such as large electrical porcelain insulators, are too big to go through tunnel kilns and they are fired in large fixed-hearth, intermittent kilns which have several control zones at various heights.

For bricks costing as little as, perhaps, 3p each, capital and running costs must obviously be kept to a minimum, but for a pottery figure costing, say, £3000 the production of a very high proportion of first-quality ware is the prime consideration.

TIME–TEMPERATURE SCHEDULING

In the firing process ware is heated to temperatures that range from 700°C for some decorated tableware to 1700°C for basic refractories over cycles of anything from a few hours to a few weeks. As shown earlier the following changes occur:

(a) removal of water;
(b) removal of binders and organic media;
(c) dehydroxylation;
(d) oxidation;
(e) decomposition;
(f) phase transformation.

KILN DESIGN CONSIDERATION

The ideal ceramic kiln would be capable of heating and cooling the ware uniformly at the maximum rate of temperature change for each of the stages mentioned above, and of controlling the maximum firing temperature. The principal points to consider in the design of a kiln are: the method of heating; the method and the materials used to construct the walls, arch and muffles; the air- and gas-flow system; the method of conveying the ware; and the temperature control system. For electric kilns, which are used in the pottery industry at temperatures up to 1300°C, all these points are covered fairly adequately in the literature.[252–257] On the other hand, the type of ware that can be fired in direct contact with the combustion products of oil and gas depends largely on the sulphur content of the fuel[258,259] which means that fuel-fired kilns can be divided into direct-fired and muffle types.

Kiln performances have improved considerably during the last 10–20 years by the use of insulating refractory brickwork,[255,260] ceramic fibre linings,[257] and the self-recuperative burner in intermittent kilns,[261,262] and by the use of oxidation-resistant silicon carbide muffle plates in muffle tunnel kilns.[263]

An air- and gas-flow system, in addition to that required for supplying air to burners and for the removal of combustion products, is desirable in all kilns to provide adequate ventilation and to assist cooling, and in tunnel kilns to heat the ware in the early stages of firing and later to cool it by counter-current heat exchange. Water cooling is seldom used in ceramics kilns apart from that supplied to the burners of high-temperature kilns in the refractories industries.

Most present-day tunnel kilns are of the car type (Figs 15 and 16) and the many attempts to introduce belt, walking-beam and pushed-bat types have only enjoyed a very limited and usually a very short-lived success. The reason is simply that a tunnel kiln demands an almost foolproof method of conveying ware through it, since any serious failure will lead to a shut-down which could disrupt production for several days. One alternative to a tunnel kiln car, that has been used successfully for almost as long, is the rotating annular platform on which ware is conveyed in circular kilns (Fig. 17). The roller-hearth kiln[264] and the Hoverkiln[256] are two current examples of tunnel kilns that do not use cars.

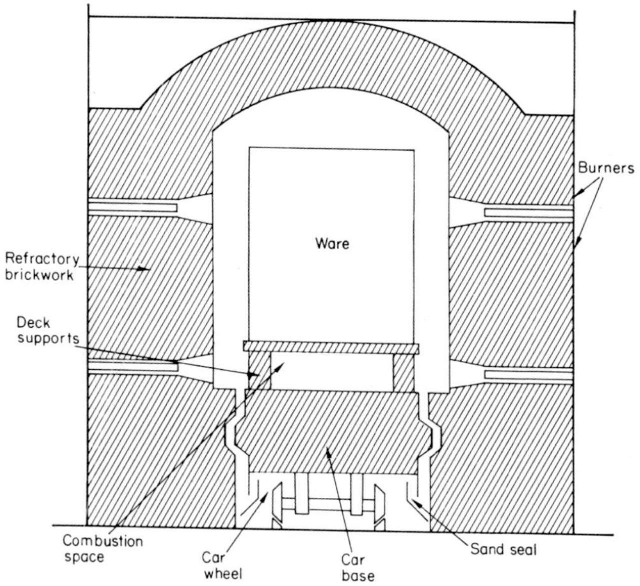

Fig. 15. Diagram of tunnel kiln (cross-section).

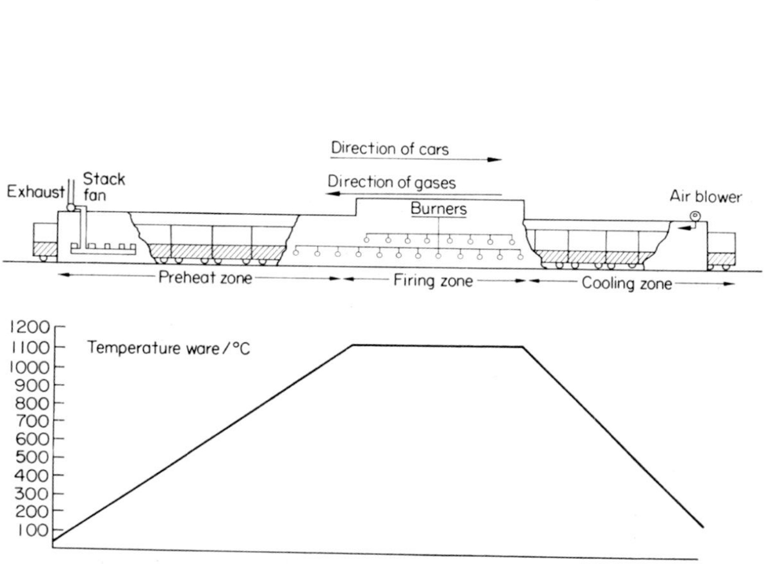

Fig. 16. Diagram of tunnel kiln (side elevation).

Fig. 17. Circular electric tunnel kiln.

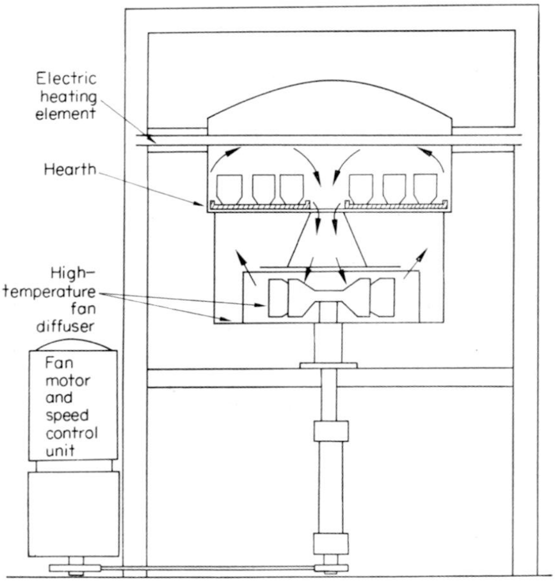

Fig. 18. Diagram of Hoverkiln.

The principle of the Hoverkiln (Fig. 18) is that air blown from a fan provides a cushion of air that can support 29 kgf m^{-2}. Cups and plates require a refractory or heat-resisting alloy bat to support them, but tiles can be supported directly on this air cushion. Although a Hoverkiln has been used to fire decorated bone china there are now no kilns of this type in use.

Both continuous and discontinuous methods of automatic temperature control are used, although the latter are much more common. Multi-zone control is very common in tunnel kilns, and in some of the more recent electric intermittent kilns for glost and decorating tableware, a three-zone control system has made it possible to exercise very close control throughout the setting. With modern intermittent kilns programme controllers are usually used so that the ware may be heated and sometimes cooled according to a predetermined schedule. Of course, not even the best control equipment can guarantee uniformity of temperature throughout the load, and temperature surveys must be made in order to obtain enough information upon which to base corrective action. Ready means of access for thermocouples and trailing leads are desirable in all tunnel kilns.

TYPES OF KILNS

Pottery Industry

For decorating and glost firing processes, which are carried out at temperatures below 1150°C, continuous and intermittent electric[252,253] or muffle kilns are used to the greatest extent. This is because element failure is not a serious problem at these temperatures and also because a clean, oxidizing atmosphere is necessary in decorating kilns and excessive air or gas movement is deleterious in glost kilns. For biscuit or once-firing processes, which are carried out between about 1150 and 1250°C, direct firing, using natural gas, propane or butane as fuel, is used in the majority of cases. Most sanitary ware is however still fired in muffle tunnel kilns, the reasons for which are largely historical, and many of these were until recently fired by oil but have now been converted to gas.

Electric and gas-fired intermittent kilns have become increasingly popular over the last 20 years or so.[254,258,265]

The types of kilns and fuels used by the pottery industry in Stoke-on-Trent in 1973 are shown in Table 10.

TABLE 10. Kilns and fuels used by the pottery industry in Stoke-on-Trent in 1973

Method of firing	*Tunnel kilns*	*Modern intermittent kilns*
Natural gas	220	235
Electricity	91	589
Fuel oil	4	—
Liquefied petroleum gases	38	10

Heavy Clay and Refractories Industries

Many continuous-chamber kilns are still in use and a few are still fired by coal. There had been a strong trend to oil firing in the 1960s[266] but conversion to firing by gas and LPG have since taken place. The fuels and method of firing down-draught intermittent kilns are given in Table 11 for the years 1961, 1967 and 1973.

TABLE 11. Fuels and methods of firing down-draught intermittent kilns in the heavy clay and refractories industries

Year	Coal Hand fired	Coal Mechanically fired	Producer Gas fired	Oil fired	Total
1961	2224	297	12	266	2799
1967	1134	231	Nil	678	2043
1973	35	60	Nil	328	423

This shows that the number of such kilns has reduced by 85% in 12 years and that the number of those being coal fired has fallen to about 4%. Producer gas firing actually ceased by 1965, whilst oil firing increased considerably and reached a peak by 1967. Since 1970 conversions to natural gas[267–269] and liquefied petroleum gases[269–271] have taken place.

There are also more than 150 modern car tunnel kilns mostly fired by oil although conversions to natural gas and liquefied petroleum gases have taken place. However, since the price increases in petroleum fuels, gas firing is gaining ground and some coal firing is again being considered.

Modern intermittent kilns are now being used in the heavy clay and refractories industries.

CAPITAL AND RUNNING COSTS

Approximate capital costs for tunnel kiln installations in the pottery, heavy clay and refractories industries are given in Table 12. The running costs, i.e. fuel, labour, kiln ancillaries and maintenance, are given in Table 13.

The approximate outputs corresponding with the costs in Tables 12 and 13 are given in Table 14.

KILN ATMOSPHERE

Almost all ceramic products manufactured in the U.K. are fired under oxidizing conditions. Hard-paste porcelain, which requires a reducing atmosphere, is manufactured at only one factory where firing is carried out in gas-fired tunnel kilns of German design. In the heavy clay industry, products such as blue bricks require a reducing atmosphere in the later stages of firing. Previously the necessary conditions were obtained in down-draught intermittent kilns by feeding the fires heavily with coal and excluding all secondary air. Unfortunately dense, black smoke is emitted from the chimneys,[272] and gas firing has now been introduced to enable this firing

TABLE 12. Approximate capital costs for tunnel kiln installations

Type of industry	*Installation*	*Cost*
Pottery	Decorating kilns	£60 000
	Glost kilns	£80 000
	Biscuit kilns	£120 000
Heavy clay and refractories	Building bricks	£300 000
	Refractory products	
	(1400°C)	£280 000
	(1750°C)	£380 000

TABLE 13. Approximate running costs for tunnel kilns

Type of industry	*Installation*	*Cost per hour*
Pottery	Decorating kilns	
	electric	£3
	gas	£2
	Glost and biscuit kilns	
	gas	£5
Heavy clay	Building bricks	
	oil	£15
Refractories	Products fired below 1500°C	
	oil	£16

TABLE 14. Outputs of tunnel kilns

Type of industry	*Installation*	*Output per hour*
Pottery	Decorating kilns	400 items
	Glost and biscuit kilns	700 items
Heavy clay	Building bricks	2000 bricks
Refractories	Products fired below 1500°C	300 bricks

process to comply with the provisions of the Clean Air Act. Contrary to popular belief the typical surface colour of blue bricks is not due to the development of ferrous silicates but to reflection from a well-developed skin of hematite (ferric oxide) crystals. It appears that the ferric oxide is derived from a liquid formed near the brick surface under reducing conditions.[273]

FUTURE DEVELOPMENTS

In the National Board for Prices and Incomes' report[274] on the pottery industry the following statement appeared:

'Considerable effort is being devoted in the industry to reducing the rigidity imposed by tunnel kilns, for example by speeding up the firing time and supplementing them with intermittent kilns, but it seems clear that tunnel kilns will dominate the industry for a good many years. Technological change in the industry since their introduction has largely centred upon them and is likely to do so in the future.'

Intermittent kilns have, nevertheless, become popular in the pottery industry because of the limitations of tunnel kilns[254] and a cost analysis[275] claims that, compared with tunnel kilns, the labour savings achieved with intermittent kilns can easily offset their higher fuel costs. The means of improving the operation of orthodox tunnel kilns has also been reviewed[254] and some of the newer fast-firing tunnel kilns described.[254]

Heat transfer considerations[276] indicate that rapid heating is possible for single articles whereas long cycles are necessary for closely stacked ware. Single-layer settings for articles such as pipes and sanitaryware are as high as those of some kilns with multi-deck settings, but wide, low settings are inevitable for the single-layer firing of tableware, tiles and bricks.[277] In a car tunnel kiln wide, low settings result in a very high car-to-ware-weight ratio, which is one of the reasons why bat (refractory slabs on which the ware is set) kilns such as the roller-hearth kiln,[264] the Trent kiln[278] and the Hoverkiln[256] have been developed. The Italian S.I.T.I. roller-hearth kiln has the advantage of layers of ware at four levels and therefore it yields four times the output for a given floor area.[254] A bat is also a much simpler and cheaper ware-carrying unit than a kiln car and loaded bats can be stored in magazines (Fig. 19) that occupy only a fraction of the floor area required for loaded kiln cars. In Sweden, for example, there are sanitaryware plants[279] with British pushed-bat kilns that have four- and five-level storage systems. With these advantages one might suppose that

Fig. 19. S.I.T.I. tile kiln.

bat kilns are replacing car tunnel kilns to a marked extent. But, in fact, there are only a few bat kilns operating in this country and one reason is that they tend to be more complicated and less dependable than car tunnel kilns.[254] Another reason is that many bat kilns are electric and have high fuel costs and element replacement problems. However, one-high, rapid firing kilns may eventually be favoured because the ease with which they can be loaded and unloaded automatically allows them to be used as hot conveyors in fully automatic production lines.

Rapid firing would also assist quality control by giving early warning of any faults that either arise or manifest themselves during the firing process. If rapid firing is adopted widely bat kilns would almost certainly become more popular, and rapid drying systems would also be required.

The changes that are now taking place in the availability and price of fuels are influencing the choice of kiln. For example, in the pottery industry sulphur-free fuels such as natural gas and liquefied petroleum gases are creating a trend away from muffle and electric kilns towards direct firing. Also even in the heavy clay and refractories industries, where oils of high sulphur content can be used in direct-fired kilns, the relative prices and future prospects of oil and gas are now such that gas firing has rapidly gained ground.

Finally, attention must be drawn to the increasing practice of building intermittent kilns on the manufacturer's premises and packaging them ready for road haulage and perhaps shipment to the customer's factory where they can be brought into use very quickly. Small tunnel kilns are also being built in sections which are packaged and transported to the customer. In addition to rapid assembly these kilns also have the advantage of being readily dismantled and reassembled or even of being modified by removing or adding sections.

TERMS USED IN THE CERAMICS INDUSTRIES

For the meanings of terms used in the ceramics industries the reader should consult *Dictionary of Ceramics* by A. E. Dodd, Newnes-Butterworths, London.

13
Equipment (Metals)

EQUIPMENT SELECTION

The choice of furnace employed for the sintering operation is influenced by:

(a) compact material, which dictates the necessary sintering time, temperature and atmosphere;
(b) component size;
(c) output requirement;
(d) capital and running costs.

In the production of engineering components, the need for consistency, combined with high output rates, has resulted in almost exclusive use of the continuous sintering furnace, on which the ensuing description will concentrate.

THE BASIC FURNACE FUNCTIONS

In general, the continuous furnace used for sintering metal parts is a tunnel-like, gas-tight assembly consisting of three basic sections: the burn-off section; the heating and soaking chamber; and the cooling section.

The Burn-Off Section

Although the process is sometimes carried out separately in air, a burn-off section, in which pressing lubricant is volatilized from green compacts prior to sintering, is more usually an integral part of the modern furnace. The size of this section and its heat input should ensure that lubricant is adequately expelled from the heaviest part at its maximum loading density. At the same time, the heating rate should not be so fast as to cause compact disruption by violent de-waxing.

Burn-off products can upset the balance between the sintering atmosphere and the work and can harmfully affect furnace furniture. Care is therefore taken to ensure that the flow of atmosphere gas is such as to prevent their ingress into the heating and cooling zones of the furnace.

The Heating and Soaking Chamber

This section is usually divided into a number of heating zones in order to facilitate the necessary degree of temperature control. The bulk of heat input is devoted to the first zone in which work, pre-heated in the burn-off section, normally attains full process temperature. Heat to other zones, in which the work soaks and full bonding is achieved, equates merely to that required to compensate for furnace losses. The length of this section is dictated by the soaking time and output required.

The Cooling Section

Cooling normally takes place in a water-jacketed chamber, the length of which should ensure that the work temperature on discharge into air is low enough to avoid oxidation of the components.

FURNACE DESIGN CONSIDERATIONS

Because of the severe environmental conditions often imposed in the sintering furnace, the methods of heating, atmosphere retention and work conveyance demand particular scrutiny here.

Heating Methods

A sintering furnace may be either fuel-fired or electrically heated. Gas-firing must be indirect, either by radiant tube or by externally heating a muffle, since products of combustion must not be allowed to impair the sintering atmosphere composition. Electric elements, however, predominate as the means of heating used in the U.K.

The life of a furnace element depends upon:

(a) operating temperature;
(b) watt-density loading;
(c) cross-sectional area;
(d) frequency of on/off cycling;
(e) method of support;
(f) working atmosphere.

Three materials are commonly employed: nickel–chromium alloys; silicon carbide elements; and molybdenum elements.

Nickel–chromium alloys

Various grades are employed in the form of sinuous tape or rod, for furnace temperatures up to 1150°C. The heat dissipation requirement, in terms of watt-density loading, should be as low as possible in order to minimize element operating temperature. A heavy cross-section is advantageous in forestalling environmental attack with its resultant effect upon electrical resistance; attack deriving from the sintering atmosphere, which can result in ultimate element failure, may take various forms[280] including:

(a) Carbon pick-up: soot deposition from the gas or from lubricant burn-off products can give rise to arcing and shorting-out effects at element bends and lead-outs. Thus it is advisable to run at low element voltages. Furthermore, in carburizing atmospheres the element materials may also absorb carbon, causing the formation of low melting point phases and ultimate failure by localized melting.

(b) 'Green rot'.
(c) Sulphur attack.
(d) Lead and zinc contamination.

Silicon carbide elements

Silicon carbide rods can be used at higher temperatures than nickel–chromium elements (up to a furnace temperature of 1400°C) and have the advantage of being easily and quickly replaced even when the furnace is still hot. Higher watt loadings can be employed but, on the debit side, silicon carbide is brittle, fragile and susceptible to thermal shock. Furthermore, there is a tendency for the material to age in service, i.e. its resistance increases with time at elevated temperatures, and a means of adjustment of voltage is normally required with equipment into which this type of element is incorporated. The use of silicon carbide resistors in conjunction with highly carburizing atmospheres is significantly less restricted than in the case of nickel–chromium elements.

Molybdenum elements

This expensive material, capable of relatively high watt-density loadings, is restricted to special high-temperature applications with an upper limit of approximately 1700°C. Care must be taken in the use of molybdenum elements because:

(a) they are highly susceptible to thermal shock;
(b) the metal forms a volatile oxide so that it is restricted to use in high-hydrogen or fully inert atmospheres;
(c) their use in carburizing atmospheres usually results in poor life.

Retention of Satisfactory Atmosphere Conditions

In the operation of a sintering furnace, it is essential that the atmosphere should remain consistent in terms of composition in order to ensure reproducibility of results with respect to part strength, carbon content, dimensional stability, etc. It is, therefore, important to guard against ingress of air into the furnace which will upset this condition. This is achieved by using either a gas-tight furnace shell lined with refractory bricks, which are in contact with the gas, or a muffle, usually manufactured from a nickel–chromium alloy.

The electrically-fired burn-off zone inevitably contains a muffle to protect the elements in that section from attack by the products of lubricant volatilization. However, the decision to use a muffle in the sintering chamber depends upon the nature of the sintering duty. The non-muffle furnace is adequate for applications such as the sintering of non-ferrous components in exothermic gas or iron-base parts of low or medium carbon content in endothermic gas. However, where a purer atmosphere is required, as in the sintering of higher carbon compacts, the muffle construction is preferred. Although a muffle adds to the capital and maintenance costs of a furnace lower dew points are attainable and atmosphere conditioning time is significantly reduced, since the gas is no longer in contact with refractory bricks bearing reducible oxides.

Method of Work Conveyance

The manner in which work is transported through a continuous sintering furnace is dictated by the loading density necessary to achieve the desired output and by the process temperature, 1150°C being normally regarded as the limit for the use of nickel–chromium alloys in mechanical conveyers. As it is traditional, in the industry, to identify the equipment in terms of the method employed to carry the work, the conveyance techniques are best considered as part of the review of sintering furnace types which follows.

TYPES OF FURNACE

Sintering Furnaces for Use up to 1150°C

The mesh belt conveyor furnace

For temperatures up to 1150°C, this furnace (shown diagrammatically in Fig. 20) is the most widely employed in the metal sintering industry. In this case, work is carried on a conveyor in the form of an endless belt, woven from nickel–chromium alloy wire, through a burn-off section, sintering chamber, which may be of the muffle or non-muffle construction, and cooling zones.

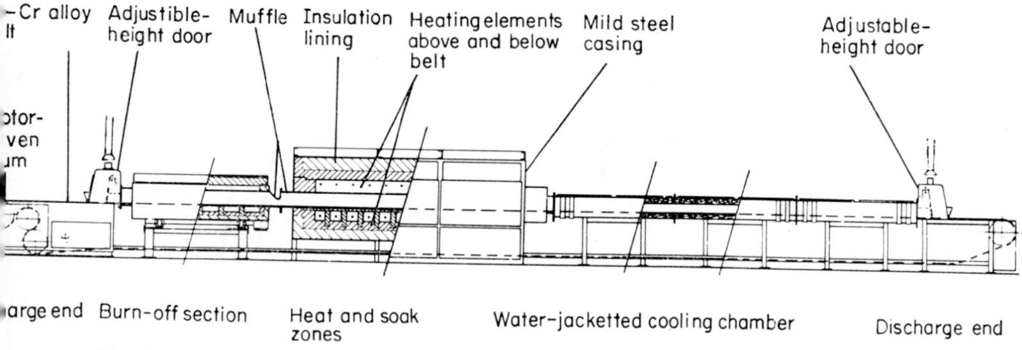

Fig. 20. Partly-sectioned view of a mesh belt conveyor sintering furnace with muffle.

The motor-driven belt, which passes over large diameter drums fitted at each end of the furnace, is normally between 12 and 24 in wide on production equipment. In typical sintering cycles, these give outputs between approximately 170 and 430 lb h^{-1} of, say, iron-base components. The loading of the furnace is limited by the strength of the belt at the sintering temperature and normally lies between 5 and 10 lb ft^{-2} of belt at temperatures employed for iron-base parts.

The straight-through furnace (Fig. 21) is usually used in conjunction with an exothermic or endothermic atmosphere. The design of the hump-back furnace (Fig. 22) aids the conservation of atmosphere used, which is normally a hydrogen–nitrogen mixture in this case.

Fig. 21. A view of two 18″ mesh belt conveyor muffle furnaces (courtesy, Birlec Ltd).

Fig. 22. A 24″ humpback mesh belt conveyor furnace (courtesy, Efco Furnaces Oxy Metal Industries (G.B.) Ltd).

The roller hearth furnace

For larger outputs and heavier components, the roller-hearth furnace may be considered an alternative to mesh belt conveyor furnaces. In this case, the work, loaded in light trays, is transported through the furnace on driven rollers which are capable of a loading density several times greater than that of the mesh belt conveyor over the same temperature range. Because the rollers are normally manufactured from nickel–chromium alloys, the maximum working temperature is again limited to 1150°C. An obvious limitation of this type of furnace is that it cannot be adopted for use with a muffle.

High-temperature Sintering Furnaces

The pusher furnace

In the pusher design (Fig. 23) a line of loaded trays is progressed through the furnace by an intermittently operating pusher gear (which may be mechanical, hydraulic or pneumatic) at the charge end. The furnace can be used below 1150°C, where it constitutes a useful compromise between the limited loading density of the mesh belt conveyor furnace and the high production rates of the roller-hearth type. However, it is primarily regarded as a furnace for high-temperature sintering.

Fig. 23. Molybdenum-element pusher furnace for sintering in reducing atmosphere at temperatures up to 1750°C (courtesy, FHD Furnaces Ltd).

In order to minimize the forces necessary to push the trays, several variations on the theme have been introduced:

(a) inclination of the furnace, where the charge end is higher than the discharge end, so that gravity helps to lessen the pushing force;
(b) the push–pull furnace, where trays are linked together to form a chain which is simultaneously pushed and pulled through the furnace, the charge-end pushing force and discharge-end pull being balanced so as to exert minimum load on the hot trays. This design is suitable for sintering temperatures up to 1350°C.

The walking beam furnace

The walking beam concept relies on two or more longitudinal beams which can move in gaps in the hearth of the furnace (Fig. 24). At a pre-set point in time, these beams are lifted to raise the charge from the hearth, they move forward a predetermined distance, lower in order to deposit the charge and then return to their original position. By intermittent repetition of this cycle of events, the charge, contained in light simple trays, progressively moves through the furnace. The beams are moved mechanically

Fig. 24. High-temperature walking beam furnace for sintering in hydrogen-bearing atmospheres (couresty, Efco Furnaces Oxy Metal Industries (G.B.) Ltd).

and are contained in a gas-tight furnace housing to permit the use of a protective atmosphere.

As a result of the more severe temperature conditions, maintenance costs for these furnaces are significantly greater than for the low-temperature types.

Batch Sintering Furnaces

Batch furnaces are employed for sintering in, essentially, three types of application:

(a) for the production of conventional work at low outputs;
(b) for special duties. Because allowances need not be made for moving parts, the batch furnace can be designed to operate at higher temperatures than continuous equipment. Furthermore, since it is possible to make seals more effectively, the batch furnace can be used in conjunction with purer atmospheres;
(c) for experimental work.

Batch furnaces, of essentially the same design as employed for conventional heat treatment, e.g. the bell, the pit and the hand pusher, have been used, mainly for conventional sintering work. Other equipment includes the direct-resistance furnace, a classical method capable of very high temperatures which is almost universally used in the manufacture of refractory metals such as tungsten (3000°C). In this technique the powder compact, surrounded by a protective atmosphere in a bell, constitutes the heating element. In the hot-pressing furnace, the die and powder are heated *in situ* under a press, heat input being by convection using external heaters, induction or direct resistance. The maximum temperature is obviously limited to the maximum useful working temperature of the available die materials and is not normally higher than 1000°C.

Probably the most widely used batch furnace is the vacuum type. Heating[280,281] may be by induction (Fig. 25), direct resistance or heating elements. Element materials used include nickel–chromium (up to 1000–1150°C depending on furnace arrangement), molybdenum (1800°C) tungsten and tantalum (2500°C). Graphite is being used up to 2200°C element temperature. In conventional plant, it is obviously important to burn off any pressing lubricant separately before sintering in vacuum in order to avoid damage to furnace equipment and pumps and contamination during sintering. Furnaces have recently been introduced in which de-waxing can be carried out in the sintering furnace, eliminating the need for intermediate handling and transfer.[282]

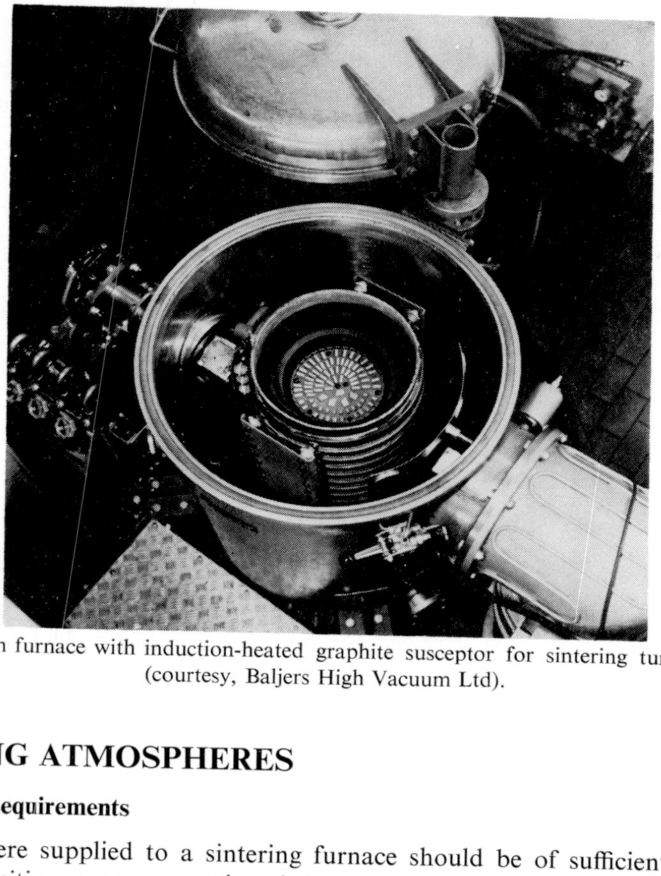

Fig. 25. Vacuum furnace with induction-heated graphite susceptor for sintering tungsten carbide (courtesy, Baljers High Vacuum Ltd).

SINTERING ATMOSPHERES

Atmosphere Requirements

The atmosphere supplied to a sintering furnace should be of sufficient volume to maintain a positive purge, preventing air ingress, and its flow so arranged as to sweep out lubricant volatiles. In terms of composition, it should satisfy the following requirements.

(a) In order to promote bonding, the sintering atmosphere must be sufficiently reducing at process temperature to reduce oxides on the metal powder or, at least, be inert in order to prevent their formation. The gas should also be capable of maintaining an oxide-free finish during cooling.

(b) In the treatment of carbon-bearing materials, the atmosphere must also possess a controllable carbon potential.

Atmosphere Types and Their Generation

Atmospheres suitable for sintering have been described elsewhere[280,283,284] and only a brief summary is attempted here. They fall broadly into three groups in ascending order of cost and sophistication:

Exothermic and endothermic gases

These atmospheres are produced by the controlled combustion of a fuel gas, such as propane, butane, natural gas or town gas, with air. Their composition is dependent upon the air-to-gas ratio and type of fuel employed. A typical compositional range is shown in Fig. 26.

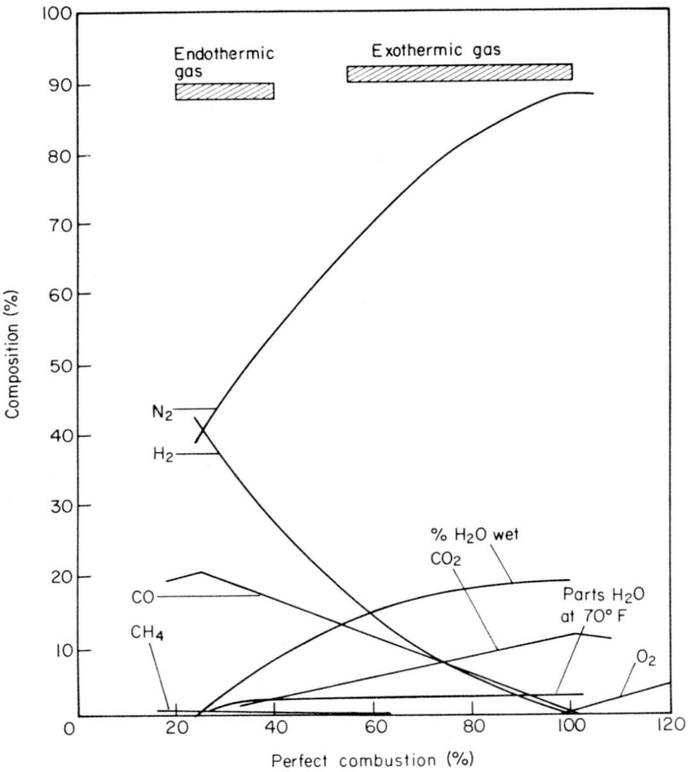

Fig. 26. Approximate guide to the composition of generated atmospheres in relation to the percentage of perfect combustion of fuel gas with air.

By virtue of its composition (i.e. relatively high CO_2 and water vapour content), exothermic gas has limited reducing power. If sufficiently rich, it is capable of preventing the oxidation of iron but is unable to prevent decarburization. Therefore, its use is restricted in ferrous sintering to low-carbon or pure iron compacts. It is most widely employed for non-ferrous work such as the copper-base compacts. The gas is produced in a generator (Fig. 27) which provides a means for burning controlled air–gas mixtures in a refractory-lined combustion chamber at temperatures in the range 1100–1450°C. The dew point of the product gas leaving the generator is approximately 5°C above the cooling water temperature. Removal of CO_2 and water vapour, to produce 'stripped and dried exothermic gas'[285] yields a more reducing atmosphere of wide carbon potential capability which constitutes an alternative to endothermic gas.

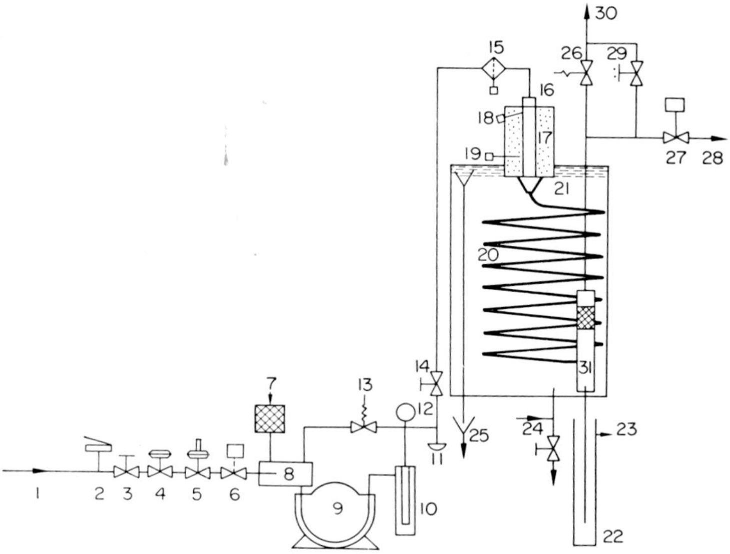

Fig. 27. Diagram of an exothermic gas generator (courtesy, Birlec Ltd). 1 Gas inlet, 2 Pressure switch, 3 Stop valve, 4 Primary governor, 5 Zero governor, 6 Solenoid valve, 7 Air inlet and filter, 8 Proportioning valve, 9 Compressor, 10 Oil and water separator, 11 Explosion disc, 12 Pressure gauge, 13 Relief by-pass, 14 Flow control valve, 15 Flame trap with heat fuse, 16 Multiport burner and igniter, 17 Combustion chamber, 18 Spy hole, 19 Flame failure probe, 20 Cooling chamber, 21 Water tank, 22 Condensate seal pot, 23 Condensate drain, 24 Water inlet and drain, 25 Water outlet, 26 Relief valve, 27 Solenoid valve, 28 Product outlet, 29 Switched vent valve, 30 Vent, 31 Water separator.

Endothermic gas contains significantly lower oxidant levels as generated than exothermic gas and is capable of a useful range of carbon potentials at normal sintering temperatures. For this reason it is widely employed in the sintering of carbon-bearing iron compacts where the avoidance of decarburization is necessary. In the endothermic gas generator a controlled rich mixture of gas and air is passed over a catalyst bed in an externally heated retort so that complete combustion is obtained. The typical gas plant, of which a simplified diagram is shown in Fig. 28, is normally capable of generating the gas at dew points in the range -15 to $+15$°C.

Hydrogen-based gases

Where higher alloy powder, such as stainless steel or hard metal, is to be sintered, a purer atmosphere of greater reducing power is demanded. This is the basis for the use of gases with higher hydrogen contents. Pure hydrogen is used in special cases only, because of its high cost. A cheaper source of hydrogen is 'cracked ammonia' derived from the thermal dissociation of anhydrous ammonia to give 75% H_2, 25%

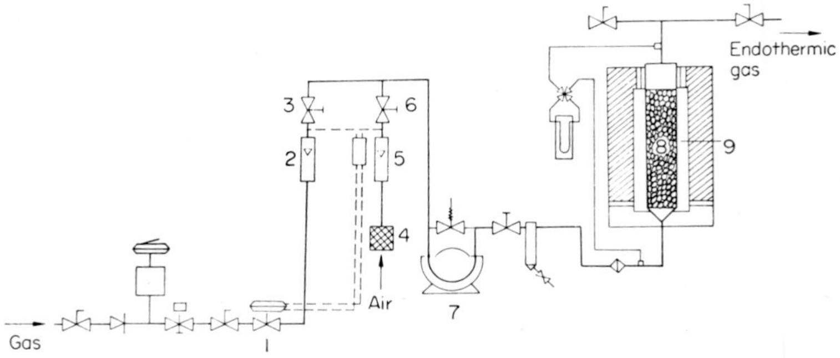

Fig. 28. Diagram of an endothermic gas generaotr (courtesy, Birlec Ltd). 1 Zero governor, 2 Gas flowmeter, 3 Gas control valve, 4 Filter, 5 Air flowmeter, 6 Air control valve, 7 Compressor, 8 Heated retort with catalyst, 9 Control thermocouple.

N_2. This is an endothermic reaction effected in practice by passing pre-vaporized ammonia through a heated retort containing a suitable catalyst (Fig. 29). Normally the dew point of cracked ammonia from the dissociator lies between -15 and -40°C depending on the quality of the ammonia feedstock. De-ammoniators can be used to reduce water vapour and residual ammonia contents further.

A cheaper derivative of ammonia, 'burned ammonia', ranges from moderately-reducing H_2–N_2 mixtures containing 25% H_2 to almost pure nitrogen.

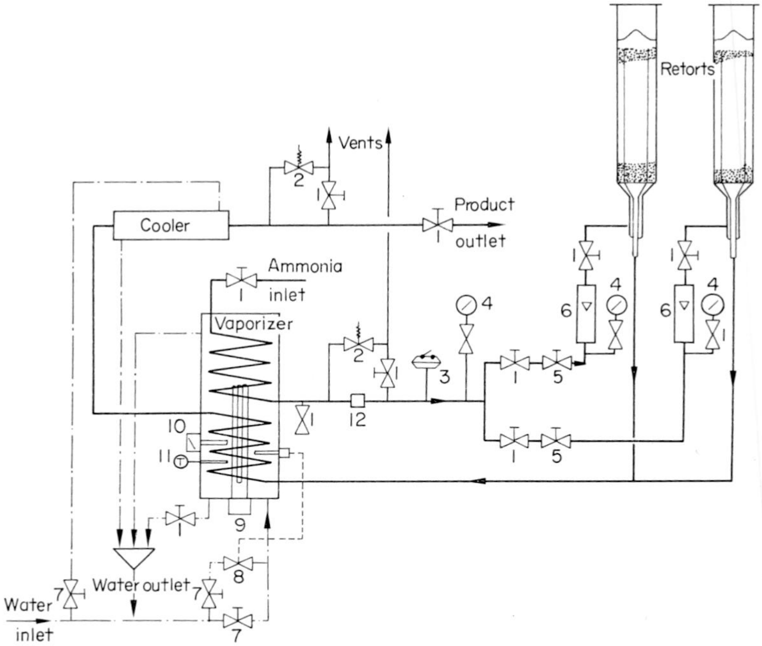

Fig. 29. Diagram of an ammonia (courtesy, Birlec Ltd). 1 Ammonia line stop valves, 2 Pressure relief valves, 3 Low-pressure switch, 4 Pressure gauges, 5 Pressure-reducing valves, 6 Flowmeters, 7 Water line isolating valves, 8 Thermostatic valve, 9 Vaporizer heaters, 10 Thermostat, 11 Thermometer, 12 Oil filter.

Other atmospheres

The use of bottled argon and helium, in batch furnaces, is restricted to special applications, such as the sintering of reactive metals, because of their high cost. Batch sintering in vacuum is fairly widely practised for such materials as stainless steel,[286] hard metals, Alnico magnets and refractory metals.

ATMOSPHERE CONTROL

The general principles of atmosphere control in the heat treatment of metals are well established[283] and its particular role in sintering has been the subject of a recent study.[287] Consideration of equilibrium data relating to oxidizing–reducing and carburizing–decarburizing reactions enables a useful guide to critical atmosphere requirements to be formulated. Diagrams of the type shown in Figs 30 and 31 are of particular use.

Several reviews of the available methods of atmosphere control have been published,[282,287] and contain detailed descriptions of techniques. Commercially, the monitoring of endothermic gas has received most attention recently in the context of carbon potential control. As indicated by Fig. 30, carbon potential is assessed by measurements of water vapour, in terms of dew point, or CO_2 content. The principal techniques used industrially are summarized below.

Dewpoint Measurement

The dew cup

This simple, classical method involves the progressive cooling of a polished cup surface enveloped in the gas under test. A mist of water droplets condenses on the surface when it attains the dew point temperature.

The fog chamber

This instrument operates on the principle that a rapidly expanding gas, cooled adiabatically, produces a condensate mist only when specific conditions of pressure drop, ambient temperature and moisture level in the sample gas are satisfied.

The lithium chloride dew cell

This instrument depends upon the behaviour of the hygroscopic salt lithium chloride when in contact with water vapour. Dry lithium chloride absorbs water at room temperature and dissolves forming a saturated solution. This solution may in turn be heated to the temperature at which the evaporation tendency of the moisture just matches the absorption tendency of the salt. This temperature is directly related to the dew point temperature.

Infrared CO_2 Analysis

When infrared radiation passes through a gaseous mixture, certain gases absorb energy at characteristic wavelengths in proportion to their concentration. This principle is utilized in the measurement of CO_2 and other compound gases but cannot be applied to elemental gases which do not absorb in the infrared spectrum. Com-

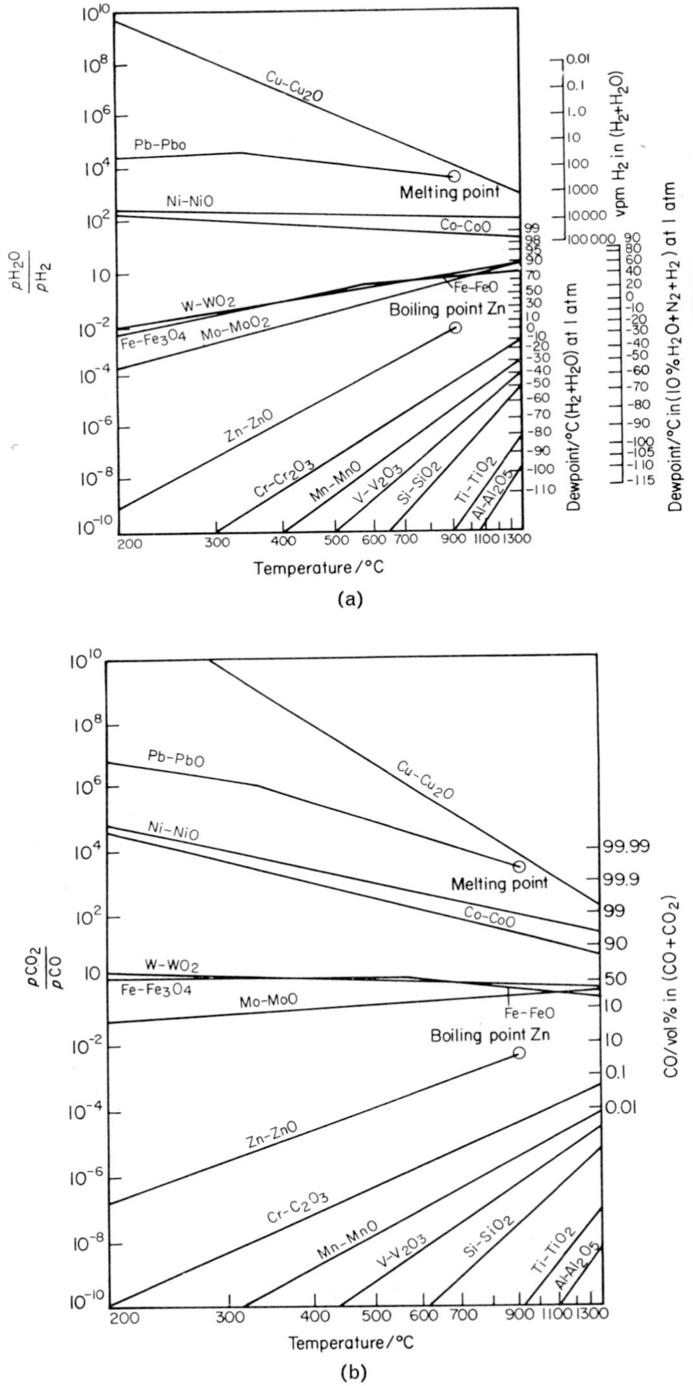

Fig. 30. Critical requirements for the oxidation of selected metals with temperature in atmospheres containing (a) water vapour and hydrogen, (b) carbon dioxide and carbon monoxide.[4]

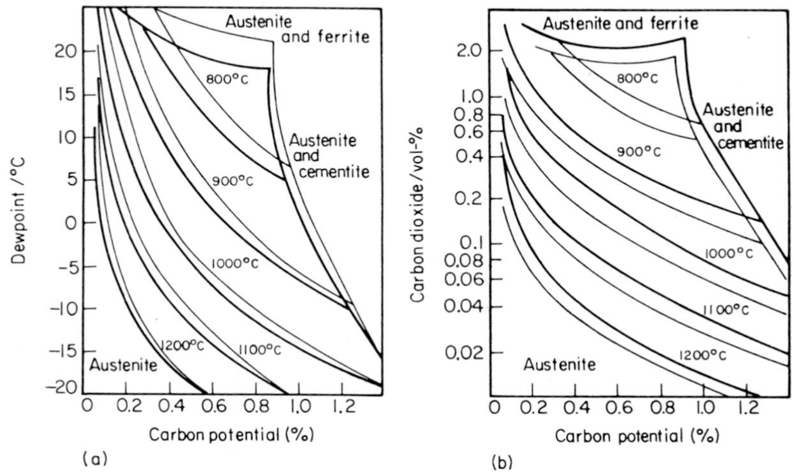

Fig. 31. Carbon potential in typical endothermic gas as measured by (a) dew point, and (b) vol. % of CO_2. The heavy curves refer to endothermic gas derived from propane; the light curves refer to endothermic gas derived from natural gas.

mercially available analysers for atmosphere control rely upon the detection of the difference in absorption level between equal pulses of infrared radiation passing through the sample gas and through a non-absorbing reference gas such as nitrogen.

Compared with dew point measurement, infrared CO_2 analysis is more sensitive to changes in atmosphere composition, more rapid in response and less susceptible to sampling problems. For these reasons, there has been a trend towards its use for automatic control of endothermic gas generators in preference to dew point measurement, although the latter is, in terms of capital cost, cheaper for single point control.

SINTERING EQUIPMENT FOR HARD METALS

As hard metals tend to be a special case and have their own particular problems they will be dealt with separately.

Two main types of furnace are now used:

(a) Continuous-stoking tube furnaces based upon molybdenum resistance heating. These furnaces invariably use a hydrogen atmosphere. Such furnaces if properly designed and operated can be satisfactory and economical. Temperature control needs constant skilled attention, and even so thermal cycles can vary with factors such as boat load and stoking speed. Atmosphere control to avoid carbon gain or loss is not easy. This is the major technical shortcoming of this type of furnace.

(b) Vacuum furnaces using either graphite resistance or high frequency heating. These are generally batch furnaces. If these are properly designed and operated temperature gradients can be virtually non-existent and temperature control can be very good. Atmosphere control can be excellent. However cobalt has a fairly high vapour pressure and care must be exercised to avoid loss of cobalt and difficulties from cobalt vapour. For this reason moderate vacuum only can be used.

Modern developments are towards continuous or semi-continuous vacuum furnaces and, especially for throwaway tip manufacture, towards a continuous de-wax, pre-sinter and final sinter furnace. Considerable progress has been made in this last field.

The difficulties that arise in the design of suitable furnaces for the hard metal industry mainly result from the very fine particle size of the powder which is used. The ball-milled carbides ready for pressing can have average grain sizes in the range 0·6–5·0 μm, but they contain many much finer particles of both carbide and cobalt. The immense surface area makes these powders very sensitive to atmospheric conditions. Substantial oxidation can occur at room temperature, to an extent which can cause serious practical difficulties. At the temperatures of up to 950°C which are used in de-waxing and presintering, oxidation–reduction and carburizing–decarburizing reactions readily occur to an extent that can seriously impair the properties of the sintered product.

The need to remove pressing lubricant (generally paraffin wax) can seriously affect this operation. Unless wax vapour is completely removed from the system in a low-temperature zone the remaining traces of vapour can seriously affect the compacts. Economic trends, and the poor green strength of hard metal compacts, make it desirable to sinter in one operation directly after pressing. This can be done in several furnaces now commercially available, but it is doubtful whether they have all completely solved the de-waxing problem.

The immense use of the indexible insert, which is not resharpened after use, has greatly increased the need for precise and accurate atmosphere control. Any reactions which occur between a compact and its surrounding atmosphere proceed most extensively at the corners of the compact where surface area-to-volume ratio is at a maximum. Yet with these inserts it is especially important that the material properties at the extreme corners are as good as possible. With continuous-stoking furnaces where the work must be packed in boats of refractory powder such as alumina, reaction with the packing material can be especially troublesome.

Hard metal alloys have properties which are extremely sensitive to grain size, yet the very fine particle size and liquid phase sintering make these alloys very sensitive to rapid grain growth in sintering. Grain growth inhibitors such as vanadium and tantalum carbides can be very effective, but are best avoided because of their embrittling effect. Thus the best solution is a very accurate control of sintering temperature and of temperature gradients within the furnace.

These requirements are not readily fulfilled. The temperatures involved, in the range 1350–1550°C, are not readily measured. Thermocouples deteriorate rapidly and optical methods require much continued skilled attention if their accuracy capabilities are to be fully realized. In continuous furnaces using a hydrogen atmosphere, temperature distribution across the furnace tube can be negligible, but longitudinal gradients can be variable with boat load and stoking speed. As temperature sensing points must be fixed this leads to considerable difficulty with temperature control.

With the batch type of vacuum furnaces serious temperature gradients can easily arise because of the very bad heat transfer from the heating elements. At low temperatures this can only occur by conduction through the furnace parts and even in the best of furnace designs severe gradients exist until, as sintering temperature is approached, heat transfer by radiation becomes very efficient. Temperature measurement by optical means can be very efficient and needs a minimum of attention with such furnaces. However, it is generally only possible to measure temperature some considerable distance from the heating elements. This can cause serious troubles from thermal lag, and control systems designed to minimise these effects are essential.

FUTURE DEVELOPMENTS

Furnaces

Although there are obvious advantages, in terms of product properties, to be gained from high-temperature sintering, the trend towards its wider commercial use appears to have lost momentum. The reasons for this are probably twofold:

(a) In general, the maintenance costs associated with high-temperature furnaces are high and future developments in this field depend largely on the introduction of reliable materials for furnace components to operate under the severe conditions imposed. For example, in the pusher furnace, the choice of tray material, able to withstand at high temperature the forces necessary to push a row of work, is critical.[288]

(b) The introduction of improved powders, either of higher compressibility or the pre-alloyed types, enables high strengths to be developed in parts sintered at conventional 'low' temperatures.

In view of (b), it is anticipated that immediate developments will concentrate on the design of furnaces operating at temperatures up to 1150°C but capable of higher output than currently attainable in order to cope with the demands of a fast-expanding industry. This objective can be achieved in two ways:

(a) by the wider use of conveyance mechanisms such as the roller-hearth, which allow higher loading densities than the mesh belt furnace which predominates in the industry today;

(b) by persevering with the mesh belt but employing stronger materials or reinforced conveyor mechanisms in order to reduce hot stresses on the belt.[289]

With larger furnaces, the saving in heating costs to be gained by the use of gas rather than electricity will assume significant proportions. Although the fuel has been used to heat burn-off sections, there is a reluctance to use gas-firing in the sintering chamber which, again, involves high maintenance costs associated with gas-fired radiant tubes and muffles at temperatures in excess of 1000°C. The development of reliable ceramic tubes[290] and muffles will be welcomed in this direction.

Larger furnaces will obviously occupy more valuable floor space. Since the major portion of a sintering furnace length consists of cooling section when employing the conventional water-jacketed chamber system, the introduction of forced-cooling techniques offers advantages.[289] The possibility of better-quality powders leading to reduced sintering time requirements and, thus, shorter furnaces is also relevant in this context.

The new sintering processes may call for different furnace concepts:

(a) The sinter–forging operation,[291–294] in which sintered metal preforms are forged to give parts of high density and final, or near final, shape, has received wide publicity recently. The heating equipment required will depend upon the manner in which the process is eventually performed. At present, three possibilities appear to exist:

(i) Forging of pre-form removed at forging temperature from the conventional type of sintering furnace, modified to suit. This technique poses difficult handling problems.
(ii) Parts forged in a separate operation after sintering. In this case, pre-heating by induction seems the likely path.
(iii) Combining sintering and heating for forging using induction.[295]

(b) As powder rolling techniques[296] reach viability, a demand for specialized sintering furnaces will exist.

Recent developments in continuous-vacuum furnaces[297] may find an outlet in sintering although the evolution of lubricant volatiles still remains a problem.

Atmospheres

Although little radical change is envisaged in atmosphere-generator design, further investigation could usefully be made into the behaviour of the sintering atmosphere in the furnace, with regard to its effect on product properties and uniformity.[287,298,299] The use of a synthetic nitrogen-based atmosphere, providing an alternative to endothermic gas for sintering, has recently been described.[300]

In atmosphere control, the high temperature electrochemical cell for oxygen concentration measurement[301] may prove to be the next refinement. This instrument, more commonly known as the oxygen probe or monitor, is, in principle, a high-temperature galvanic oxygen concentration cell in which two electronically conducting electrodes, usually platinum, are in contact with the inner and outer surfaces of a solid electrolyte (zirconia) in the form of a sealed tube. If the condition of pure ionic conduction within the electrolyte is fulfilled, then a potential difference will be set up between the two electrodes, the magnitude of which is dependent upon the ratio of the oxygen partial pressures of the gases on the inside and outside of the tube, e.g. air and the furnace atmosphere respectively. This voltage is related to the oxygen potential of the atmosphere from which the critical ratios of its oxidizing–reducing and decarburizing–carburizing constituents can be inferred. Thus the probe gives, *in situ*, a more meaningful assessment of atmosphere condition than either simple CO_2 or dew point measurement.

Eventually this form of atmosphere analysis, which is relatively cheap, eliminates gas sampling problems and can be coordinated into a simple atmosphere control system, may well prove applicable to sintering atmospheres if the problem of attack from atmosphere contaminants can be overcome.

References

1. C. G. Goetzel, *Treatise on Powder Metallurgy*, Interscience, (now) Chichester, Sussex, 1949.
2. H. Hausner, in *New Methods for the Consolidation of Metal Powders* (H. Hantrer, K. Roll and P. Johnson, Eds.), Plenum, New York, 1967.
3. P. R. Marshall, *Metall. Rev.* **13**, 3.
4. H. Hausner, *Developments in Powder Metallurgy*, Plenum, New York, 1971.
5. G. H. Gessinger and M. J. Bomford, *Int. Metall. Rev.* **19**, 51 (1974).
6. C. S. Smith, *Metall. Rev.* **9**, 1 (1964).
7. F. N. Rhines, *Plansee Proceedings 1958* (F. Benesovsky, Ed.), Metallwerk Plansee, Reutte/Tyrol, p. 38.
8. D. W. Budworth, *Trans. Br. Ceram. Soc.* **69**, 29 (1970).
9. G. W. Greenwood and A. Boltax, *J. Nucl. Mater.* **5**, 234 (1962).
10. C. Zener, see for example D. McLean, *Grain Boundaries in Metals*, Clarendon Press, Oxford, 1957, p. 239.
11. N. A. Haroun and D. W. Budworth, *J. Mater. Sci.* **3**, 326 (1968).
12. W. D. Kingery and B. Francois, *J. Am. Ceram. Soc.* **48**, 546 (1965).
13. F. A. Nichols, *J. Appl. Phys.* **37**, 4599 (1966).
14. G. W. Greenwood, *Acta Met.* **4**, 243 (1956).
15. C. Wagner, *Z. Elektrochem.* **65**, 581 (1961).
16. I. M. Lifshitz and V. V. Slezov, *J. Phys. Chem. Solids* **19**, 35 (1961).
17. G. C. Kuczynski, *Trans. A.I.M.E.*, **185**, 169 (1949).
18. C. Herring. *J. Appl. Phys.* **21**, 301 (1950).
19. R. L. Coble and J. E. Burke, *Prog. Ceram. Sci.* **3**, 1971 (1963).
20. F. V. Lenel, G. S. Ansell and R. C. Morris, in *Modern Developments in Powder Metallurgy*, Vol. 4 (H. Hausner, Ed.), Plenum, New York, 1971, p. 199.
21. B. H. Alexander and R. W. Balluffi, *J. Met.* **2**, 1219 (1950).
22. R. L. Coble, *J. Appl. Phys.* **32**, 787 (1961).
23. D. L. Johnson, *J. Am. Ceram. Soc.* **53**, 574 (1970).
24. J. K. Mackenzie and R. Shuttleworth, *Proc. Phys. Soc.* **B62**, 833 (1949).
25. M. F. Ashby, *Acta Met.* **22**, 275 (1974).
26. H. S. Cannon and F. V. Lenel, *Plansee Proceedings 1952*, Metallwerk Plansee, Reutte/Tyrol, 1952, p. 106.
27. W. D. Kingery, *J. Appl. Phys.* **30**, 301 (1959).
28. D. Buist, B. Jackson, I. M. Stephenson, W. F. Ford and J. White, *Trans. Br. Ceram. Soc.* **64**, 173 (1965).
29. I. M. Stephenson and J. White, *Trans. Br. Ceram. Soc.* **66**, 443 (1967).
30. J. W. Cahn and R. B. Heady, *J. Am. Ceram. Soc.* **53**, 406 (1970).
31. T. J. Whalen and M. Humenik, in *Sintering and Related Phenomena* (G. C. Kuczynski, Ed.), Gordon and Breach, New York, 1967, p. 715.
32. W. J. Huppman, in *Sintering and Catalysis* (G. C. Kuczynski, Ed.), Plenum, New York, 1975, p. 359.
33. British Patent 1 177 279.
34. R. L. Coble, *J. Appl. Phys.* **32**, 793 (1961).
35. P. J. Jorgensen and W. G. Schmidt, *J. Am. Ceram. Soc.* **53**, 24 (1970).

36. D. W. Budworth, *Min. Mag.* **37**, 833 (1970).
37. D. W. Budworth, *Phys. Sintering*, Special Issue, 119 (1971).
38. C. E. Hage and J. A. Pask, *Phys. Sintering* **5**, 109 (1973).
39. W. Beere, *Acta Met.* **23**, 131 (1975).
40. P. Murray, E. P. Rodgers and A. E. Williams, *Trans. Br. Ceram. Soc.* **53**, 474 (1954).
41. T. Vasilos, *J. Am. Ceram. Soc.* **43**, 517 (1960).
42. G. E. Mangsen, W. A. Lambertson and B. Best, *J. Am. Ceram. Soc.* **43**, 55 (1960).
43. R. E. Jaeger and L. Egerton, *J. Am. Ceram. Soc.* **45**, 209 (1962).
44. J. D. McClelland, *J. Am. Ceram. Soc.* **44**, 526 (1961).
45. T. Vasilos and R. M. Spriggs, *J. Am. Ceram. Soc.* **46**, 493 (1963).
46. M. S. Koval'chenko and G. V. Samsonov, *Poroshk. Metall.* **1**, 3 (1961).
47. C. Herring. *J. Appl. Phys.* **21**, 437 (1950).
48. S. Scholz and B. Lersmacher, *Arch. Eisenhüttenwesen*, **41**, 98 (1964).
49. R. C. Rossi and R. M. Fulrath, *J. Am. Ceram. Soc.* **48**, 558 (1965).
50. S. Scholz, *Special Ceramics 1962*, Academic Press, London, 1963, p. 293.
51. F. R. N. Nabarro, *Report of a Conference on the Strength of Solids*, Physical Society, London, 1948, p. 75.
52. G. M. Fryer, *Trans. Br. Ceram. Soc.* **66**, 127 (1967).
53. G. M. Fryer, *Trans. Br. Ceram. Soc.* **68**, 185 (1969).
54. J. Williams, *Proc. Br. Ceram. Soc.* **3**, 1 (1965).
55. D. I. Matkin, *Sci. Ceram.* **5**, 441 (1970).
56. E. R. Stover, *A Critical Survey of Characterization of Particulate Ceramic Raw Materials* AFML-TR-67-56, 1967.
57. W. H. Gitzen, *Alumina Ceramics*, AFML-TR-66-13, 1966.
58. I. B. Cutler, in *High Temperature Oxides*, Part III (A. M. Alper, Ed.), Academic Press, London, 1970, p. 124.
59. D. T. Livey, in *High Temperature Oxides*, Part III (A. M. Alper, Ed.), Academic Press, London, 1970, p. 124.
60. R. Thompson, *Borides: Their Chemistry and Applications*, The Royal Institute of Chemistry, London, 1965.
61. J. G. M. de Lau, *Bull. Am. Ceram. Soc.* **49**, 572 (1970).
62. D. W. Johnson and F. J. Schnettler, *J. Am. Ceram. Soc.* **53**, 440 (1970).
63. R. M. Dell, *Chem. Ind.* 1101, (1970).
64. J. L. Woodhead, *Sci. Ceram.* **4**, 105 (1968).
65. D. E. Nietz, R. B. Bennett and M. J. Synder, Office of Naval Research, AD 712817, Oct. 1970.
66. M. G. Harwood, G. L. MacDonald and V. J. Middel, *Proc. Br. Ceram. Soc.* **3**, 49 (1965).
67. I. E. Denton, D. I. Matkin and N. A. Hill, *Proc. Br. Ceram. Soc.* **12**, 33 (1969).
68. T. P. Meloy, *Ultra-Fine Grain Ceramics* (J. J. Burke, N. C. Reid and V. Weiss, Eds.), p. 17 (Syracuse University Press, 1970).
69. L. D. Hart and L. K. Hudson, *Bull. Am. Ceram. Soc.* **43**, 13 (1964).
70. J. S. Hirschhorn, *Powder Tech* **4**, 1 (1970/71).
71. W. C. Chiu and D. E. Day, presented at 72nd Annual Meeting of the American Ceramic Society, Philadelphia, May 1970 (5-R-70).
72. D. Lewis and M. W. Lindley, *J. Am. Ceram. Soc.* **49**, 49 (1966).
73. E. J. Wheeler and D. Lewis, *J. Mater. Sci.* **4**, 681 (1969).
74. Mamoru Senna and Hiroshi Kuno, *J. Am. Ceram. Soc.* **54**, 259 (1971).
75. L. D. Hart, *Bull. Am. Ceram. Soc.* **43**, 26 (1964).
76. M. D. Ashton and F. H. H. Valentine, *Trans. Inst. Chem. Eng.* **44**, 166 (1966).
77. *Ceram. Age*, **87**, 69 (1971).
78. M. L. Roessler and H. C. Willis, *Bull. Am. Ceram. Soc.* **48**, 284 (1969).
79. R. C. Rossi, R. M. Fulrath and D. W. Fuerstenan, *Bull. Am. Ceram. Soc.* **49**, 289 (1970).
80. D. W. Fuerstenan and Jahan Fouladi, *Bull. Am. Ceram. Soc.* **46**, 821 (1967).
81. M. O. Warman and D. W. Budworth, *Trans. Br. Ceram. Soc.* **66**, 253 (1967).
82. C. A. Bruch, *Ceram. Age*, **83**, 44, (1967).
83. N. Claussen and J. Jahn, *Powder Metall. Int.* **2**, 87 (1970).
84. *Ceramic Fabrication Processes* (W. D. Kingery, Ed.), MIT, Cambridge, Mass., 1958.

85. M. A. Strivens, *Bull. Am. Ceram. Soc.* **42**, 13 (1963).
86. Institution of Chemical Engineers Working Party on 'Consolidation of Particulate Matter', Task Unit on 'Mechanics of Granulation'.
87. H. J. Helsing, *Powder Metall. Int.* **1**, 62 (1969).
88. P. C. Kapur and D. W. Fuerstenan, *Trans. AIME*, **229**, 348 (1964).
89. R. L. Brown and J. C. Richards, *Principles of Powder Mechanics*, Pergamon Press, Oxford, 1970.
90. H. Rumpf, in, *Agglomeration* (W. A. Knapper, Ed.), Interscience Publishers, (now) Chichester, Sussex, 1962.
91. P. C. Kapur and D. W. Fuerstenan, *J. Am. Ceram. Soc.* **50**, 14 (1967).
92. J. H. Grimes and K. T. B. Scott, *Powder Metall.* **11**, 213 (1968).
93. J. M. Fletcher and C. J. Hardy, *Chem. Ind.* 48 (1968).
94. J. P. Cuer, J. Elston and S. J. Teichner, *Bull. Soc. Chem.* 81, 89 and 94 (1961).
95. T. I. Barry, R. K. Bayliss and L. A. Lay, *J. Mater. Sci.* **3**, 229 (1968).
96. K. S. Mazdiyasni, C. T. Lynch and J. S. Smith, *J. Am. Ceram. Soc.* **60**, 532 (1967).
97. D. A. Everest, I. G. Sayce and B. Selton, *J. Mater. Sci.* **6**, 218 (1971).
98. D. I. Matkin, T. M. Valentine, W. Munro and J. A. Desport, Reactive hot-pressing of fine-grain alumina from vapour-deposited powders. *Proc. Br. Ceram. Soc.* **20**, 133 (1972).
99. W. E. Kuhn, (Ed.) *Ultrafine Particles*, Wiley, New York, 1962.
100. J. Hinnuber and W. Kinna, *Hartmetalltechnik und Forschung Tech. Mitt Krupp*, **19**, 130 (1961).
101. H. E. Fischmeister and H. E. Exner, Beobachturgen uber den Mahllvorgang bes Hartmetall-pulvern. *Planseeber Pulvermetall.* **13**, 178 (1965).
102. J. Hinnuber, O. Rudiger and W. Kinna, An electron microscope and X-ray investigation of the milling of tungsten carbide–cobalt mixtures. *Powder Metall.* **8**, 1 (1961).
103. E. Lardner, Review of Current Hard Metal Technology, Iron and Steel Institute, London. *Materials for Metal Cutting*, Special Publication 1970, p. 126.
104. P. B. Anderson, Hard metals of increased toughness. *Planseeber Pulvermetall.* **15**, 180 (1967).
105. H. E. Exner and J. Gurland, A review of parameters influencing some mechanical properties of tungsten carbide–cobalt alloys. *Powder Metall.* **13**, 13 (1970).
106. H. E. Exner and J. Gurland, Guide to literature on tungsten carbide–cobalt alloys. *Powder Metall. Int.* **2**, 59, 104 (1970).
107. H. F. Fischmeister and H. E. Exner, *Planseeber Pulvermetall.* **13**, 178 (1965).
108. V. A. Ivensen, N. B. Baranova, S. S. Loseva, I. G. Smatalova and O. N. Edyk, Investigation of the wet milling of hard alloy mixtures. *Tverdye Splavy*, **1**, 7 (1958).
109. C. N. Walley, private communication.
110. L. R. Blair, A study of decalcomania blistering. *Bull. Am. Ceram. Soc.* **27**, 301 (1948).
111. J. A. Ferguson, Loss of strength during dehydroxylation of clays. *J. Aust. Ceram. Soc.* **3**, 1 (1967).
112. W. H. Holmes, R. W. Cox, E. Davies and E. Rowden, Pink centres and black cores. *Trans. Br. Ceram. Soc.* **61**, 571 (1962).
113. J. R. Schorr and J. O. Everhart, Thermal behaviour of pyrite and its relation to carbon and sulphur oxidation in clays. *J. Am. Ceram. Soc.* **52**, 351 (1969).
114. P. Fisher, D. Harkort and A. W. Norris, *The Bloating of Vitrified Ceramic Bodies*, IXth International Ceramic Congress, Brussels, 1964.
115. E. M. Levin, C. R. Robbins and H. F. McMurdie, *Phase Diagrams for Ceramists*, American Ceramic Society, Columbus, Ohio, 1964.
116. W. H. Holmes, The firing of clay based ceramics. *Sci. Prog.* (*London*), **60**, 237 (1972).
117. D. I. Matkin, I. E. Denton, T. M. Valentine and P. Warrington, The fabrication of silicon nitride by ceramic/plastic technology, *Proc. Br. Ceram. Soc.* **22**, 291 (1973).
118. D. I. Matkin, I. E. Denton, T. M. Valentine and P. Warrington, The fabrication of silicon nitride by ceramic/plastic technology. *Proc. Br. Soc. Ceram.* **22**, 291 (1973).
119. C. W. Forrest and P. Kennedy, The fabrication of Refel silicon carbide components by isostatic pressing. *Spec. Ceram.* **6**, 183 (1975).
120. R. L. Brown, D. J. Godfrey, M. W. Lindley and E. R. W. May. Advances in the technology of silicon nitride ceramic. *Spec. Ceram.* **5**, 345 (1972).
121. P. E. D. Morgan and E. Scala, The formation of fully dense oxides by hydroxides. *Sintering and Related Phenomena* (G. C. Kuczynski, N. A. Hooton and C. Gibbon, Eds.), Gordon and Breach, London, 1967, pp. 861–894.

122. T. A. Wheat and T. G. Carruthers, The hot pressing of magnesium hydroxide and magnesium carbonate. *Science of Ceramics*, Vol. 4 (G. H. Stewart, ed.), Academic Press, London, 1968, pp. 33–52.
123. A. C. D. Chaklader and L. G. McKenzie, Reactive hot pressing of clays and alumina. *J. Am. Ceram. Soc.* **49**, 477 (1966).
124. T. G. Carruthers and T. A. Wheat, Hot pressing of kaolin and of mixtures of alumina and silica. *Proc. Br. Ceram. Soc.* **3**, 259 (1965).
125. C. G. Harmon, Forming clay ware by hot pressing makes new products possible. *Brick Clay Rec.* **94**, 15, 46 (1939).
126. P. Popper, Reaction sintering with special reference to non-oxide ceramics. *Trans. 7th Int. Ceramic Congress*, 1960, pp. 451–460.
127. M. C. Regan and J. W. Isaacs, The reaction sintering behaviour of compacted U+C powders. *Spec. Ceram.* 11 (1963).
128. M. C. Regan and J. Williams, The sintering of U–C–Fe alloys in the presence of a liquid phase. *Powder Metall.* **8**, 135 (1961).
129. R. K. Stringer and L. S. Williams, Reaction pressing, a new fabrication concept for intermetallic compounds, *Spec. Ceram.* **4**, 37 (1968).
130. R. W. Gooding and N. J. Parratt, Solid titanium nitride and other refractory compounds made by direct gas/metal reaction, *Powder Metall.* **11**, 42 (1963).
131. N. L. Parr, G. F. Martin and E. R. W. May, Preparation, micro-structure and mechanical properties of silicon nitride, in *Spec. Ceram.* 102 (1960).
132. N. L. Parr, R. Sands, P. L. Pratt, E. R. W. May, C. R. Shakespeare and D. S. Thompson, Structural aspects of silicon nitride, *Powder Metall.* 152 (1961).
133. C. F. Cooper, C. M. George and L. Winter, Aluminium nitride crucibles: raw materials preparation, characterization and fabrication, *Spec. Ceram.* **4**, 1 (1968).
134. P. Popper and D. G. S. Davies, The preparation and properties of self-bonded silicon carbide, *Powder Metall.* **8**, 411 (1961).
135. A. Hendry and K. H. Jack, The preparation of silicon nitride from silica, *Spec. Ceram.* **6**, 199 (1975).
136. C. W. Forrest, P. Kennedy and J. V. Shennan, The fabrication and properties of self-bonded silicon carbide bodies. *Spec. Ceram.* **5**, 99 (1972).
137. P. Popper and S. N. Ruddlesden, The preparation, properties and structure of silicon nitride, *Trans. Br. Ceram. Soc.* **60**, 603 (1961).
138. A. Atkinson, P. J. Leatt and A. J. Moulson, The role of nitrogen flow into the nitriding compact in the production of reaction-sintered silicon nitride, *Proc. Br. Ceram. Soc.* **22**, 253 (1973).
139. D. R. Messier and K. Wong, Kinetics of formation and mechanical properties of reaction-formed silicon nitride. Nov. 1973. Proc. 2nd Army Materials Technological Conf., Hyannis, Mass., U.S.A. (Eds. J. J. Burke *et al.*, Brook Hill Pub. Co. 1974.)
140. I. Amato, D. Martorana and M. Ross, The influence of raw material and process variables on reaction-bonded silicon nitride. *Powder Metall.* **18**, 339 (1975).
141. D. J. Godfrey, The effects of impurities, additions and surface preparation on the strength of silicon nitride. *Proc. Brit. Ceram. Soc.* **25**, 330 (1975).
142. D. S. Thompson and P. L. Pratt, The structure of silicon nitride, in *The Science of Ceramics*, Vol. 3 (G. H. Stewart, Ed.), Academic Press, London, pp. 33–57.
143. D. Cratchley, *Met. Rev.* **10**, 37, 79 (1965).
144. T. A. Greening, AFPL-TR-67-181, April 1967.
145. J. C. Withers and E. F. Abrams, *Plating (Paris)* **55**, 605 (1968).
146. A. A. Baker, S. Harris and E. Holmes, *Met. Mater.* **1**, 211 (1967).
147. A. A. Baker, M. B. P. Allery and S. J. Harris, *J. Mater. Sci.* **4**, 242 (1969).
148. R. G. C. Arridge and D. Heywood, *Br. J. Appl. Phys.* **18**, 447 (1967).
149. D. Cratchley and A. A. Baker, *Metal. ABM* **69**, 153 (1964).
150. P. W. Jackson, A. A. Baker, D. Cratchley and P. J. Walker, *Powder Metall.* **11**, 1 (1968).
151. A. L. Hoffmanner, Contract No. NOW 65-0281-P Final Rept., August 1966, ER-6545-7.
152. A. Toy, *J. Mater.* **3**, 43 (1968).
153. V. L. Godwin and M. Herman, February 1969, AD 853869.

154. F. Galasso, M. Salkind, D. Kuehl and V. Patarini, *Trans. Metall. Soc. AIME*, **236**, 1748 (1966).
155. K. G. Kreider, R. D. Schile, E. M. Breinan and M. Marciano, July 1968, AFML-TR-68-119.
156. J. A. Alexander, R. G. Shaver and J. C. Withers, July 1966, NASA CR-523.
157. E. M. Breinan and K. G. Kreider, *Met. Eng. Q.* **9**, 5 (1969).
158. A. G. Metcalfe and G. K. Schmitz, Paper from Soc. of Automotive Engineers Conf., October 1967, Paper 670862.
159. A. L. Cunningham and J. A. Alexander, 12th SAMPE Symposium 1968, Paper AC 15.
160. A. E. Vidoz, J. L. Camahort and F. W. Crossman, *J. Compos. Mater.* **3**, 254 (1969).
161. P. W. Jackson and J. R. Marjoram, *J. Mater. Sci.* **5**, 9 (1970).
162. F. S. Galasso and J. Pinto, *Fibre Sci. Technol.* **2**, 303 (1970).
163. P. W. Jackson, unpublished work.
164. A. A. Baker, C. Shipman, P. A. Cripwell and P. W. Jackson, *Fibre Sci. Technol.* **5**, 285 (1972).
165. R. T. Pepper, J. W. Upp, R. C. Rossi and E. G. Kendall, *Metall. Trans.* **2**, 117 (1971).
166. A. A. Baker, A. Martin and R. J. Bach, *Compos. Mater.* **2**, 129 (1971).
167. P. W. Jackson, D. M. Braddick and P. J. Walker, *Fibre Sci. Technol.* **5**, 219 (1972).
168. A. A. Baker, C. Shipman and P. W. Jackson, *Fibre Sci. Technol.* **5**, 213 (1972).
169. A. Kelly and J. Davies, *Metall. Rev.* **10**, 37 (1965).
170. N. J. Parratt, *Chem. Eng. Prog.* **62**, 3 (1966).
171. R. G. Schierding and Oliver De S. Deex, *J. Compos. Mater.* **3**, 618 (1969).
172. H. Hahn, A. P. Divecha, P. J. Lare and R. A. Hermann, ASME Conf. on Materials Technology 20–24 May, 1968, p. 231.
173. U.S. Patent 3 525 610, August 1970.
174. M. J. Kuderko, *Micrograin Solids.* Proc. 1st Int. Cemented Carbide Conf. Feb. 1971, S.M.E. EM71-937.
175. O. Kasukawa, *Cutting Characteristics and Practical Merits of Micrograin Tungsten Carbides*, Proc. 1st Int. Cemented Carbide Conf. Feb. 1971, S.M.E. EM71-938.
176. E. R. Almdale, *Carbide Wear Applications*, Proc. 1st Int. Cemented Carbide Conf. Feb. 1971, S.M.E. EM71–908.
177. W. D. Jones, *Fundamental Principles of Powder Metallurgy*, Edward Arnold, London, 1960, pp. 806–819.
178. J. R. Chalkley and D. M. Thomas, The Tribological Aspects of the Metal Bonded Diamond Wheels. *Powder Metall.* **12**, 24 (1969).
179. S. L. Hoyt, *Trans. Am. Inst. Min. Metall. Eng., Inst. Metals Div.* **89**, 9 (1930).
180. *The German Hard Metal Industry*, BIOS Final Report 1945 No. 1385 Item No. 21, pp. 41–52. HMSO.
181. P. Murray, D. T. Livey and J. Williams, in *Ceramic Fabrication Processes* (W. D. Kingery, Ed.), Wiley, New York, 1958, p. 141.
182. A. R. Hall and W. Watt, Some experiments on the hot-pressing of zirconium carbide powder at 2000°C. Selected Government Research Reports 1952. Vol. 10, *Ceramics and Glass*, HMSO, London, p. 91.
183. E. Roeder and S. Scholz, A simple hot-press for laboratory investigations. *Spec. Ceram.* 269 (1965).
184. J. G. Campbell and A. R. Ford, Graphite for hot-pressing dies. *J. Br. Ceram. Soc.* **2**, 68 (1965).
185. R. M. Spriggs, L. A. Brisette, M. Rossetti and T. Vasilos, Hot-pressing ceramics in alumina dies. *Am. Ceram. Soc. Bull.* **42**, 477 (1963).
186. G. H. Haertling, Hot-Pressed Lead Zirconate-Titanate Ceramic Containing Bismuth. *Am. Ceram. Soc. Bull.* **43**, 875 (1964).
187. Deutsche Edelstahlwerke Aktiengesellschaft, Br. Pat. 899 915, 'Hot-Pressing Die', 1962.
188. H. Moss and W. P. Stollar, Die Design for Pressure-Sintering. *Am. Ceram. Soc. Bull.* **45**, 792 (1966).
189. G. T. Moors, Br. Pat. 899 183, 'Improvements relating to Hot-Pressing Apparatus'. A.E.I. Ltd. 1962.
190. R. A. J. Sambell, 'The Technology of Ceramic Fibre-Ceramic Matrix Composities'. *Composites*, **1**, 276 (1970).
191. N. I. Naiguz and D. S. Mil'shtein, A Press for Hot-Pressing Parts from Powders of High Melting Point Materials. *Sov. Powder Metall. Ceram.* (Eng. Trans.) **3**, 206 (1962).

192. R. E. Johnson, Hot-Pressing High-Density Small Grain-Size Beryllia. *Am. Ceram. Soc. Bull.* **43**, 886 (1964).
193. G. H. Zehms and J. D. McClelland, Semicontinuous Hot-Pressing. *Am. Ceram. Soc. Bull.* **42**, 10 (1963).
194. E. G. Wolff, Laboratory Vacuum Hot-Press to 2500°C. *Powder Metall.* No. 11, 93 (1963).
195. J. A. Desport, A Technique for Ceramic Encapsulation of Thermocouples Using a Glow Discharge Electron Beam. 1971 U.K.A.E.A. Report No. AERE-R6807 (to be published in *J. Nucl. Eng.*).
196. A. B. Auskern and W. G. Thompson, Temperature Indication During Hot-Pressing. *Am. Ceram. Soc. Bull.* **44**, 459 (1965).
197. E. S. Hodge, Elevated Temperature Compaction of Metals and Ceramics by Gas Pressures. *Powder Metall.* **7**, 168 (1964).
198. W. M. Long and P. Snowdon, Isostatic Hot-Pressing: Practical Experience at AWRE Aldermaston. *Powder Metall.* **12**, 209 (1969).
199. M. J. Ryan and C. A. MacMillan, Advances in Hot-Isostatic Pressing. *Batelle Tech. Rev.* **17**, 15 (1968).
200. J. B. Huffadine, A. J. Whitehead and M. J. Latimer, A Simple Hot Isostatic Pressing Technique. *Proc. Br. Ceram. Soc.* 1969 No. 12, 'Fabrication Science, 2', p. 201.
201. L. Egerton and C. A. Bieling, Isostatically hot-pressed sodium-potassium niobate transducer material. *Am. Ceram. Soc. Bull.* **47**, 1151 (1968).
202. G. D. Barbaras, U.S. Pat. 3 455 682, 'Isostatic Hot-Pressing of Refractory Bodies', E. I. Du Pont de Nemours, 1969.
203. W. A. Pfeiler and C. K. Valentine, U.S. Pat. 3 419 935, 'Hot Isostatic Pressing Apparatus', USAEC, 1968.
204. R. P. Levey and A. E. Smith, U.S. Pat. 3 363 037, 'High Temperature Isostatic Pressing Apparatus', 1968.
205. G. J. Oudemans, 'A Continuous Hot-Pressing Technique', *ibid*, Ref. 24, p. 83.
206. R. W. Rice, in *Ultra-Fine Grain Ceramics*, J. J. Burke, N. L. Reed and V. Weiss (Eds.), Syracuse University Press, Syracuse, N.Y., 1970.
207. R. M. Spriggs, L. Atteras and R. B. Runk, Thermochemical processing of ceramics, *ibid*, Ref. 24, p. 65.
208. R. W. Leonard, Direct explosive compaction of powder materials. *Batelle Tech. Rev.* **17**, No. 10 (1968).
209. S. W. Porembka, Explosive Compaction, *Ceram. Age*, **79**, 69 (1963).
210. R. L. Hallse, High-energy forming of glasses and ceramics. *Am. Ceram. Soc. Bull.* **42**, 711 (1963).
211. T. H. Hall, 'Ultra High-Pressure, High Temperature Apparatus; the "Belt"'. *Rev. Sci. Instrum.* 1960, Vol. 31, No. 2, p. 125.
212. W. B. Daniels and M. T. Jones, Simple apparatus for the generation of pressure above 100 000 atmospheres. Simultaneously with Temperatures above 3000°C. *Rev. Sci. Instrum.* **32**, 885 (1961).
213. A. J. Delai, R. M. Haag, R. M. Spriggs and T. Vasilos, Quarterly Report No. 1. AF 19 (628)-2943.
214. D. Kalish and E. V. Clougherty, High-pressure hot-pressing refractory materials. *Am. Ceram. Soc. Bull.* **48**, 570 (1969).
215. C. M. Schwartz, Attainment of High Pressures at High Temperatures, in *High-Temperature Materials and Technology* (I. E. Campbell and E. M. Sherwood, Eds.), Wiley, New York, 1967, p. 674.
216. R. W. Rice, The Effect of Gaseous Impurities on the Hot-Pressing and Behaviour of MgO, CaO and Al_2O_3 *ibid*, Ref. 24, p. 99.
217. G. D. Miles, R. A. J. Sambell, J. Rutherford and G. W. Stephenson, Fabrication of Fully Dense Polycrystalline Magnesia, *Trans. Br. Ceram. Soc.* **66**, 319 (1967).
218. R. A. Alliegro and B. D. Foster, 'Automated Hot-Pressing of Boron Carbide Armour Plate'. 71st Ann. Meeting of the Am. Ceram. Soc., 1969.
219. G. W. Meadows, 'Hot-Pressing Process and Apparatus'. Br. Pat. 1 166 779, E. I. Du Pont de Nemours, 1969.

220. R. W. Rice, 'Production of Transparent MgO at Moderate Temperatures and Pressures', 64th Ann. Meeting, Am. Ceram. Soc., 1962.
221. R. F. Stoops, 'Liquid Phase Extrusion Hot-Pressing'. *Am. Ceram. Soc. Bull.* **48**, 225 (1969).
222. A. C. D. Chaklader, 'Transformation Plasticity in Ceramic Systems and Reactive Hot-Pressing'. *Proc. Br. Ceram. Soc.* 1970, No. 15.
223. P. E. D. Morgan and E. Scala, Report No. 365, Materials Science Centre, Cornell Univ. 1965.
224. P. E. D. Morgan and N. C. Schaeffer, 'Chemically Activated Pressure Sintering of Oxides', AFML-TR-66-356, 1966.
225. A. C. D. Chaklader and V. T. Baker, 'Reactive Hot-Pressing and Densification of Non-stabilised ZrO_2'. *Am. Ceram. Soc. Bull.* **44**, 258 (1965).
226. A. C. D. Chaklader and R. C. Cook, Kinetics of Reactive Hot-Pressing of Clays and Hydroxides. *Am. Ceram. Soc. Bull.* **47**, 712 (1968).
227. D. I. Matkin, W. Munro and T. M. Valentine, The fabrication of α-alumina by reaction hot-pressing, *J. Mater. Sci.* **6**, 7 (1971).
228. T. G. Carruthers and B. Scott, Reactive Hot-Pressing of Kaolinites. *Trans. Br. Ceram. Soc.* **47**, 185 (1968).
229. P. Stephens and E. W. Hoyt, Reactive hot-pressing of ZrC–UC solid solution'. *Am. Ceram. Soc. Bull.* **40**, 320 (1961).
230. A. Accary and R. Caillat, Study of Mechanism of Reaction Hot-Pressing. *J. Am. Ceram. Soc.* **45**, 347 (1962).
231. S. Okamoto, M. Arase and S. Okamoto, Reactive Hot-Pressing of $Ni_{1/3}Fe_{2/3}$ oxyhydroxides. *J. Am. Ceram. Soc.* **52**, 110 (1969).
232. A. Squire, *Trans. AIME* **171**, 473, 485 (1947).
233. J. Williams, Symposium on Powder Metallurgy, 1954 (Special Rept, No. 58) 112, 1956 London. (Iron and Steel Institute.)
234. J. S. Jackson, 'Hot Pressing High Temperature Compounds', *Powder Metall.* No. 8, 73 (1961).
235. E. Lardner and D. J. Bettle, *Met. Mater.* **7**, 540 (1973).
236. S. Amberg, E. P. Nylander and B. Uhrenius, *Powder Metall. Int.* **6**, 178 (1975).
237. S. Amberg and H. Doxner, Metals Society Powder Metallurgy Group Meeting, 3–4 Nov. 1975, Stratford-upon-Avon.
238. H. Suzuki, K. Hayashi, Y. Yamomoto and K. Mujake, *Jnl. Jap. Soc. Powder Metall.* **21**, 222 (1975).
239. H. Hausner, *Powder Metallurgy Forging—A Process Evaluation and Bibliography*, Metal Sciences Group, The Franklin Institute Research Laboratories.
240. P. Loewenstein, L. R. Aronin and A. L. Geary, in *Powder Metallurgy* (W. Leszynski, Ed.), Interscience, New York 563 (1961).
241. G. D. Elliott, *J.I.S.I.* Special Report 30, 1944.
242. Crook and Thompson, *J.I.S.I.* **1**, 171 (1928).
243. P. J. Zorena, M. O. Halowaty and C. M. Squarcy, *Superiority of Sinter, Blast Furnace and Steel Plant*, 1960, No. 40, 443, 451.
244. J. Dartnell, *J.I.S.I.* 1969, 207, 282. For English and Japanese Practice.
245. A. Grieve, Hot strength of sinter. *J.I.S.I.* **175**, 1 (1953).
246. R. Linder, Hot strength of sinter and breakdown. *J.I.S.I.* **189**, 233 (1958).
247. T. B. Beaden, Breakdown Without Abrasion. *J.I.S.I.* **201**, 913 (1963).
248. J. Davidson, Testing. *J.I.S.I.* **211**, 106 (1973).
249. D. F. Ball, J. Dartnell, J. Davison, A. Grieve and R. Wild, *Agglomeration of Iron Ores*, Heinemann, London, 1973.
250. W. H. Holmes, The Pottery Industry's Contribution to Clean Air, Proceedings of the National Society for Clean Air Conference, Eastbourne, October 1969, *Ceramics* **20**, 23 (1969).
251. Ninety-fifth to One Hundred and Sixth Annual Reports on Alkali, etc. 1958–1969.
252. W. H. Holmes, in *Electric Heat in the Pottery Industry, Electric Furnaces.* (C. A. Otto Ed.), Newnes, London, 1958.
253. W. H. Holmes, Review of electric firing in the pottery industry. *Claycraft*, **36**, 184 (1961).
254. W. H. Holmes, Present and future trends in firing. *J. Br. Ceram. Soc.* **4**, 289 (1967).
255. W. H. Holmes in *The Design of Walls for Intermittent Furnaces Using High Temperature Insulation.* (S. R. Probert and D. R. Hub, Eds.), Elsevier, London, 1968.

256. Anon. Introducing the Hover Kiln. *Ceramics* **18**, 44 (1968).
257. Intermittent kilns, 'Fibre Sandwich Construction'. *The Financial Times*, 11 December 1975.
258. W. H. Holmes and A. W. Norris, 'Developments in Ceramic Firing', Proceedings of Brussels Conference 1964, Institute of Fuel and Royal Society of Belgian Engineers and Industrialists.
259. E. Rowden and W. H. Holmes, 'Choice of Fuel', The Combustion Engineering Association Conference 1964. Document 7856.
260. Anon. The kiln of the future with fuel efficiency and increased output. *Claycraft* **28**, 742 (1955).
261. R. M. Davies and B. Oeppen, *J. Inst. Fuel* **45**, 383 (1972).
262. D. J. Bryan, J. Masters and R. J. Webb, Applications of self-recuperative burners, Proc. Institution of Gas Engineers. Autumn Meeting, November 1974.
263. B. H. Bates and W. Krieger, Improving the Capacity of Muffle Tunnel Kilns. *Sprechsaal* **103**, 269 (1970).
264. G. B. Margola, *Interceram.* **17**, 113 (1968).
265. E. Rowden and W. H. Holmes, Kilns and Fuels in the Ceramic Industry. *Ind. Process Heat.* November 1965.
266. E. Rowden and W. H. Holmes, Kilns and Fuels in the Ceramic Industry. *Ind. Process Heat.* October 1965.
267. Anon. 'Recent Clayworks Conversions'. *Claycraft Struct. Ceram.* **43**, 20 (1970).
268. Anon. *Br. Clayworker* **79**, 30 (1970).
269. A. E. Aldersley, The firing of heavy clay products with gas. *J. Br. Ceram. Soc.* **7**, 13 (1970).
270. D. I. Thompson, *Claycraft Struct. Ceram.* **43**, 17 (1970).
271. A. Moore, Round downdraught kilns, conversion to gas firing. *J. Inst. Fuel* **43**, 355 (1970).
272. E. Rowden, Outstanding air pollution problems in the heavy clay and refractories industries. Proceedings of the National Society for Clean Air Conference, Eastbourne, October 1969, *Br. Clayworker* **78**, 36 (1969).
273. E. W. Roberts, N. O. Riley and E. Rowden, private communication.
274. National Board for Prices and Incomes. Report 149, July 1970, Cmnd 4411, HMSO.
275. G. R. Remmey, Cost analysis, bell type or tunnel kiln. *Ceram. Age* **86**, 32 (1970).
276. W. H. Holmes, Why rapid firing is possible, *J. Br. Ceram. Soc.* **6**, 19 (1969).
277. R. F. Tatsall, *Ceram. Age* **84**, 20 (1968).
278. K. Davis and K. Manual, *Interceram.* **17**, 113 (1967).
279. R. Askelid, Automated sanitaryware production. *Ceram. Ind.* **93**, 47 (1969).
280. Powder Metallurgy Equipment Manual. Part 1 'Sintering furnaces and atmospheres' Powder Metallurgy Equipment Association (Division of the Metal Powder Industries Federation), New York, 1963.
281. O. Winkler and R. Bakish, *Vacuum Metallurgy*, Elsevier, Amsterdam, 1971.
282. J. F. Hallemeier, 'A process for delubrication, presintering, sintering and rapid cooling in a vacuum induction furnace'. *J. Vac. Sci. Technol.* **9**, 1360 (1972).
283. L. H. Fairbank and L. G. W. Palethorpe, Controlled atmospheres for the heat treatment of metals. *Heat Treatment of Metals.* Special Report 95, pp. 57–69. The Iron and Steel Institute, London, 1966.
284. *Furnace Atmospheres and Carbon Control.* ASM Monograph. American Society for Metals, Ohio, 1965.
285. J. Perry, Latest developments in controlled atmospheres—artificial atmospheres with special reference to nitrogen-based atmospheres. Part 2 of paper B1 presented at the Metal Heat Treatment Conference (Heatex), Birmingham, June, 1965.
286. G. Otto, Vacuum sintering of stainless steel. *Powder Metall. Powder Technol.* **11**, 19 (1975).
287. I. M. Coult and A. J. E. Munro, The control of exothermic and endothermic gases in sintering furnaces., *Powder Metall.* **13**, 295 (1970).
288. *Höganäs Iron Powder Handbook*, Vol. 1. Höganäs Billesholms AB, Sweden 1957–59.
289. Private communication with Drever (U.K.) Ltd.
290. J. M. Foreman and F. Y. Wong, *The Gas Council Midlands Research Station, External Report No. 149*, July 1970.
291. J. W. Lennox, Developing powder metallurgy. *Engineering* (*London*), Aug. 14th, 179 (1970).
292. G. T. Brown, The powder-forging process: a review of the basic concept and development prospects. *Powder Metall.* **14**, 124 (1971).

293. J. S. Hirschhom and R. D. Bargainier, The forging of powder metallurgy preforms. *J. Met.* **22**, 21 (1970).

294. P. J. Mullins, Powder forging—is the breakthrough near? *Iron Age Metalwork. Int.* **14**, 27 (1975).

295. J. S. Hirschhom, M. Samat and G. M. Maxwell, Induction sintering has potential for powder metal parts. *Met. Prog.* **97**, 135–136, 138 (1970).

296. I. Davies, W. M. Gibbon and A. G. Harris, Thin steel strip from powder. *Powder Metall.* **11**, 295 (1968).

297. W. C. Diman, Surveying trends in furnace technology. *Met. Prog.* **101**, 58 (1972).

298. P. R. Marshall, Experience with plant in the production of sinterings. *Powder Metall.* **9**, 163 (1966).

299. P. R. Marshall, The production of powder-metallurgy parts. *Metall. Rev.* **13**, 53 (1968).

300. A. Cook, Nitrogen-based carbon controlled atmosphere—an alternative to endothermic gas, *Heat Treat. Met.* **3**, 15 (1976).

301. L. H. Fairbank, The application of free energy–temperature diagrams and high temperature electrochemical cells in the field of furnace atmospheres. *Metallurgia* **79**, 179 (1969).

Index

A

Activated sintering, 1, 26
  additions for, 26
Additives, 18, 19
Alloying, 19
Alumina-activated sintering, 26
Alumina comminution, 18
Alumina die material, 51
Alumina hot pressing, 28
Alumina pressure-assisted sintering, 14, 15
Alumina reaction sintering, 28, 29
Alumina sintering additives, 19
Alumina whiskers, 33
Aluminium nitride
  as a die material, 51
  reaction sintered, 29, 31
Aluminium reinforcement with
  alumina whiskers, 42
  beryllium, 39
  boron, 39, 40
  borsic, 39
  carbon, 41
  silica, 34
  silicon carbide, 39, 40
  silicon carbide whiskers, 42
Ammonia, 87
Atmosphere generators, 86
  for sintering ceramics, 74
  for sintering metals, 86

B

Ball milling, 21, 22
Bat kiln, 76, 77
Batch furnaces, 84
Beryllia comminution, 18
Beryllia pressure-assisted sintering, 14
Beryllium fibres, 33
Beryllium wire reinforcement, 39
Binders, 19, 24
  removal of, 24, 25
Bingham flow, 9, 13
Blenders, 19
Blending, 19, 22
Bloating, 7
Boron fibres, 33, 39, 40
Boron phosphide, reaction sintering of, 29
Borsic fibres, 39
Bulk diffusion, 8, 9
Burn-off section, 78

C

Calcination,
  Al_2O_3, 17
  MgO, 17
  BeO, 17
Carbon fibres, 33, 40, 41
Carbon potential of atmospheres, 91
Carbon refractories, 2
Carbon silicide, *see* Silicon carbide
Chemical precipitation of powders, 17
Chemical preparation of powders, 17
Chemical vapour deposition, 41, 42
Chromium nitride, reaction sintering of, 29
Circular kilns, 71
Clay, hot pressing, 28
Comminution, 17, 18
Condensation, 9, 20
Contamination in grinding, 18
Continuous-chamber kilns, 68
  firing of, 74
Continuous pressure-assisted sintering, 54
Continuous-stoking tube furnaces, 91, 92
Copper and its alloys, sintering of, 44
  atmospheres for, 86
Co-precipitation, 17
Counter-current heat exchange, 70
Cracked ammonia, 87
Creep, 15
Crystalline strain, 18
Crystoballite, 25

D

Decomposition of carbonates, 25
Defects, of hard metals, 63

Dehydroxilation, 25
Densification, kinetics of, 9
De-waxing, 45, 78
  of carbides, 92
Dewpoint, 87, 88
  measurement of, 89
Die materials for pressure-assisted sintering,
  of ceramics, 50
  of metals, 59, 60
Dies built up, 51
Dihedral angle, 5, 6, 7, 10
Dimensional changes, 28, 46, 47
Discontinuous fibre reinforcement, 41
Dry grinding, 18
Drying ceramics, 23
Dwight Lloyd sinter strand, 65

E

Electrodeposition, 34
Evaporation, 8, 9, 20
Exothermic and endothermic gases, 86
Exothermic reactions, 28, 29
Extrusion, 42, 53, 65

F

Fibre breakage, 34, 40, 41, 42
Fibre strength degradation, 40
Filament winding, 33, 34, 39
Firing, 24, 25
Firing schedules, 24, 70
Flow properties, 19, 20
Fluid-bed drying, 19
Foil bonding, 34, 39, 40
Forging, 53
Forming processes for ceramics, 23
Freeze drying, 20
Furnace atmospheres, 80, 85
Furnace design, 79
Furnace element, life of, 79
Furnace,
  firing of, 68, 73, 74
  for pressure-assisted sintering, 52
  forced cooling, of, 93
  functions, 78
  selection of, 69, 78
  types, 68

G

Gel preparation, 20
Geometry of sintering, 4
Glost firing, 73
Grain boundary energy, 5
Grain boundary pinning, 7, 11
Grain growth, 7, 8, 11, 92
Grain growth inhibitors, 48, 92
Grain-size hard metals, 92
Granulation, 19, 21 22
Graphite as a die material, 51
Green density, 46
Greenwalt pan, 65
Grinding aids, 18
Grinding contamination, 18

H

Hard metals liquid-phase sintering, 2, 47
Hard metals pressure-assisted sintering, 59
Hard metals sintering equipment, 41
Heat transfer, 76
High-energy rate forming, 53
Hot isostatic pressing,
  ceramics, 52
  hard metals, 60
  metals, 61, 62
Hot pressing, 36, 41, 42, 52, 58, 84
Hoverkiln, 71, 73, 76
Hydrogen-based gases, 87

I

Impellor blenders, 19
Impregnated diamond tools, 49
Indexable inserts, 92
Infrared CO_2 analysis, 89
Injection moulding, 34
Intermittent kilns, 72, 76
Iron base parts sintering, 44
  atmospheres for, 86
Iron ore, 2, 65
Iron oxide comminution, 18
Isostatic pressure-assisted sintering, 52
  for hard metals, 61

K

Kaolinite, 25
Kiln atmosphere, 73, 74
Kiln, bat, *See* Bat kiln
Kiln design, 70
Kiln development, 76
Kiln firing,
  heavy clay and refractories, 74
  pottery, 73
Kiln selection, 69

Kinetics
of densification, 9
of reaction sintering, 31

L

Liquid-phase sintering, 2, 10, 47
Lubricants, 19, 22
removal of, 45, 78

M

Magnesia-activated hot pressing, 28
Magnesia-activated pressure-assisted sintering, 14, 15
Magnesia-activated reaction sintering, 28
Magnesia-activated sintering, 26
Material transport, 8
Materials variables (metals), 45
Mechanical property improvement, 57
Mechanical fibre reinforcement, 35
Mechanical hot isostatic pressing, 62
Mechanical powder forging, 64
Mesh belt conveyor furnaces, 44, 81
Metal-bonded diamond tools, 49
Metal–Metal oxide equilibria, 90
Metal borides, 17
Metallic carbides, pressure-assisted sintering of, 15
Micrograin cemented carbides, 48
Milling liquid additions, 19, 22
Molybdenum furnace elements, 80
Muffle furnaces, 73, 80
Mullite, 25

N

Neck growth, 6, 8, 9, 10
Nickel–chromium furnace elements, 79
Nitrogen-based atmospheres, 94

O

Organic-based binders, 19
Organic-based lubricants, 19
Oxidation, 25
Oxygen concentration measurement, 94

P

Packing spheres, 11
Paraffin wax, 22
composition of, 36
Particle shape, 45
Particle size, 10, 25, 45
Particle structure, 46
Phase transformation
on grinding, 18
on sintering, 25
Plasma spraying, 34, 40
Plastic flow, 8, 9
Plasticizers, 19
Plutonium oxide, activated sintering of, 26
Pores, 1, 3, 6, 13, 21, 61
gas in, 7, 11, 46
shrinkage of, 9
Potassium niobate, pressure-assisted sintering of, 14
Powder,
characteristics of, 16, 20
extrusion of, 65
forging of, 64, 94
preparation of, 16–20
porosity effects of, 21
rolling of, 94
Presintering (carbides), 92
Presses for pressure-assisted sintering, 52, 60
Press forging, 53
Pressing defects, 21
Pressure-assisted die materials, 51
Pressure-assisted reaction sintering, 28
Pressure-assisted sintering, 2, 13
Programme controllers, temperature, 73
Pusher furnace, 83

R

Radiant tubes, 93
Reaction pressure-assisted sintering, 55
Reaction sintering, 2, 17, 27
Reinforcing techniques, 34
Repetitive pressure-assisted sintering, 54
Roller-hearth furnace, 82
Roller-hearth kiln, 71, 76
Rolling, 53

S

Saggers, 24
Setting, 24
Shape changes, 28, 47
Sieving, 19
Silica,
fibre reinforcement with, 33, 34
pressure-assisted sintering of, 15
Silica allotropes, 25

Silicon carbide
  fibres, 33, 39, 40
  furnace elements, 80
  reaction sintering, 28, 29, 31, 32
  whiskers, 33
Silicon nitride, 17
  reaction sintering, 29, 31, 32
  whiskers, 33
Sinter, 2, 65
Sinter forging, *see* Powder forging
Sinter tests, 66
Sintering
  definition of, 4
  diagrams of, 10
  stages of, 6
  temperature of, 44
  time of, 45
Sintering atmospheres, 85
Sodium niobate, pressure-assisted sintering of, 14
Sol gel calcination, 20
Spray drying, 19, 20, 22
Spray roasting, 20
Strain in ground powders, 18
Surface diffusion, 8, 9
Surface energy, 4
Surface-set diamond tools, 49

T

Temperature control, 73, 92
Temperature measurement, 52
Testing, 2
Thermosetting resins, 19
Thoria, hot pressing of, 28
Titanium reinforcement
  with beryllium, 39, 40
  with boron, 39
  with borsic, 39
  with silicon carbide, 39
Titanium nitride reaction sintering, 29, 31
Transient liquid-phase pressure-assisted sintering, 55
Tumble blenders, 19
Tungsten carbide, pressure-assisted sintering of, 50
Tungsten–copper bonding of diamond, 49
Tunnel kilns, 69, 70, 75, 76

U

Ultrafine tungsten carbide products, 48
Uranium carbide,
  activated sintering of, 26
  reaction sintering, 28, 29
Uranium oxide, activated sintering, 26

V

Vacuum furnace, 8, 59
  continuous, 94
Vacuum pressure-assisted sintering, 54
Vapour-phase reaction, 20
Vibrational blenders, 19
Viscous flow, 8, 9
Viscous-phase pressure-assisted sintering, 54

W

Walking beam furnace, 83
Warp sheet lamination, 33, 34, 39, 41
Water removal in firing, 24
Wet grinding, 18
Whiskers, 42

Z

Zinc oxide, activated sintering, 26
Zone pressure-assisted sintering, 53